OPERATIONS RESEARCH: VOLUM

Student Solutions Manual
for
Winston and Venkataramanan's
Introduction to Mathematical Programming
Fourth Edition

Wayne L. Winston
Indiana University

Munirpallam Venkataramanan
Indiana University

THOMSON
BROOKS/COLE

Australia • Canada • Mexico • Singapore • Spain • United Kingdom • United States

Printed in Canada
1 2 3 4 5 6 7 07 06 05 04 03

Printer: Webcom

ISBN: 0-534-39903-7

For more information about our products, contact us at:
Thomson Learning Academic Resource Center
1-800-423-0563

For permission to use material from this text, contact us by:
Phone: 1-800-730-2214
Fax: 1-800-731-2215
Web: http://www.thomsonrights.com

Asia
Thomson Learning
5 Shenton Way #01-01
UIC Building
Singapore 068808

Australia/New Zealand
Thomson Learning
102 Dodds Street
Southbank, Victoria 3006
Australia

Canada
Nelson
1120 Birchmount Road
Toronto, Ontario M1K 5G4
Canada

Europe/Middle East/South Africa
Thomson Learning
High Holborn House
50/51 Bedford Row
London WC1R 4LR
United Kingdom

Latin America
Thomson Learning
Seneca, 53
Colonia Polanco
11560 Mexico D.F.
Mexico

Spain/Portugal
Paraninfo
Calle/Magallanes, 25
28015 Madrid, Spain

Table of Contents

Chapter 2 Solutions

Section 2.1

1a. -A= $\begin{bmatrix} -1 & -2 & -3 \\ -4 & -5 & -6 \\ -7 & -8 & -9 \end{bmatrix}$

1b. 3A= $\begin{bmatrix} 3 & 6 & 9 \\ 12 & 15 & 18 \\ 21 & 24 & 27 \end{bmatrix}$

1c. A+2B is undefined.

1d. A^T= $\begin{bmatrix} 1 & 4 & 7 \\ 2 & 5 & 8 \\ 3 & 6 & 9 \end{bmatrix}$

1e. B^T= $\begin{bmatrix} 1 & 0 & 1 \\ 2 & -1 & 2 \end{bmatrix}$

1f. AB= $\begin{bmatrix} 4 & 6 \\ 10 & 15 \\ 16 & 24 \end{bmatrix}$

1g. BA is undefined.

3. Let A = (a_{ij}), B = (b_{ij}), and C = (c_{ij}). We must show that A(BC) = (AB)C. The i-j'th element of A(BC) is given by

$$\sum_x a_{ix}\left(\sum_y b_{xy}c_{yj}\right) = \sum_x \sum_y a_{ix}b_{xy}c_{yj}$$

The i-j'th element of (AB)C is given by

$$\sum_y \left(\sum_x a_{ix}b_{xy}\right)c_{yj} = \sum_y \sum_x a_{ix}b_{xy}c_{yj} = \sum_x \sum_y a_{ix}b_{xy}c_{yj}$$

Thus A(BC) = (AB)C.

Section 2.2

1. $\begin{bmatrix} 1 & -1 \\ 2 & 1 \\ 1 & 3 \end{bmatrix} \begin{bmatrix} x_1 \\ x_2 \end{bmatrix} = \begin{bmatrix} 4 \\ 6 \\ 8 \end{bmatrix}$ and $\begin{bmatrix} 1 & -1 & 4 \\ 2 & 1 & 6 \\ 1 & 3 & 8 \end{bmatrix}$

Section 2.3

1. $\left[\begin{array}{cccc|c} 1 & 1 & 0 & 1 & 3 \\ 0 & 1 & 1 & 0 & 4 \\ 1 & 2 & 1 & 1 & 8 \end{array}\right] \left[\begin{array}{cccc|c} 1 & 1 & 0 & 1 & 3 \\ 0 & 1 & 1 & 0 & 4 \\ 0 & 1 & 1 & 0 & 5 \end{array}\right] \left[\begin{array}{cccc|c} 1 & 0 & -1 & 1 & -1 \\ 0 & 1 & 1 & 0 & 4 \\ 0 & 0 & 0 & 0 & 1 \end{array}\right]$

The last row of the last matrix indicates that the original system has no solution.

2. $\left[\begin{array}{ccc|c} 1 & 1 & 1 & 4 \\ 1 & 2 & 0 & 6 \end{array}\right] \left[\begin{array}{ccc|c} 1 & 1 & 1 & 4 \\ 0 & 1 & -1 & 2 \end{array}\right] \left[\begin{array}{ccc|c} 1 & 0 & 2 & 2 \\ 0 & 1 & -1 & 2 \end{array}\right]$

This system has an infinite number of solutions of the form
$x_3 = k$, $x_1 = 2 - 2k$, $x_2 = 2 + k$.

3. $\left[\begin{array}{cc|c}1 & 1 & 1\\2 & 1 & 3\\3 & 2 & 4\end{array}\right]\left[\begin{array}{cc|c}1 & 1 & 1\\0 & -1 & 1\\3 & 2 & 4\end{array}\right]\left[\begin{array}{cc|c}1 & 1 & 1\\0 & -1 & 1\\0 & -1 & 1\end{array}\right]\left[\begin{array}{cc|c}1 & 1 & 1\\0 & 1 & -1\\0 & -1 & 1\end{array}\right]\left[\begin{array}{cc|c}1 & 0 & 2\\0 & 1 & -1\\0 & 0 & 0\end{array}\right]$

This system has the unique solution $x_1 = 2$ $x_2 = -1$.

Section 2.4

. $\begin{bmatrix}1 & 0 & 1\\1 & 2 & 1\\2 & 2 & 2\end{bmatrix}\begin{bmatrix}1 & 0 & 1\\0 & 2 & 0\\2 & 2 & 2\end{bmatrix}\begin{bmatrix}1 & 0 & 1\\0 & 2 & 0\\0 & 2 & 0\end{bmatrix}\begin{bmatrix}1 & 0 & 1\\0 & 1 & 0\\0 & 2 & 0\end{bmatrix}\begin{bmatrix}1 & 0 & 1\\0 & 1 & 0\\0 & 0 & 0\end{bmatrix}$

Row of 0's indicates that V is linearly dependent.

2. $\begin{bmatrix}2 & 1 & 0\\1 & 2 & 0\\3 & 3 & 1\end{bmatrix}\begin{bmatrix}1 & 1/2 & 0\\1 & 2 & 0\\3 & 3 & 1\end{bmatrix}\begin{bmatrix}1 & 1/2 & 0\\0 & 3/2 & 0\\3 & 3 & 1\end{bmatrix}\begin{bmatrix}1 & 1/2 & 0\\0 & 3/2 & 0\\0 & 3/2 & 1\end{bmatrix}$

$\begin{bmatrix}1 & 1/2 & 0\\0 & 1 & 0\\0 & 3/2 & 1\end{bmatrix}$

$$\begin{bmatrix}1 & 0 & 0\\0 & 1 & 0\\0 & 3/2 & 1\end{bmatrix}\begin{bmatrix}1 & 0 & 0\\0 & 1 & 0\\0 & 0 & 1\end{bmatrix}$$

The rank of the last matrix is 3, so V is linearly independent.

Section 2.5

1. $\left[\begin{array}{cc|cc}1 & 3 & 1 & 0\\2 & 5 & 0 & 1\end{array}\right]\left[\begin{array}{cc|cc}1 & 3 & 1 & 0\\0 & -1 & -2 & 1\end{array}\right]\left[\begin{array}{cc|cc}1 & 3 & 1 & 0\\0 & 1 & 2 & -1\end{array}\right]\left[\begin{array}{cc|cc}1 & 0 & -5 & 3\\0 & 1 & 2 & -1\end{array}\right]$

Thus $A^{-1} = \begin{bmatrix} -5 & 3 \\ 2 & -1 \end{bmatrix}$

3. $\left[\begin{array}{ccc|ccc} 1 & 0 & 1 & 1 & 0 & 0 \\ 1 & 1 & 1 & 0 & 1 & 0 \\ 2 & 1 & 2 & 0 & 0 & 1 \end{array}\right]$ $\left[\begin{array}{ccc|ccc} 1 & 0 & 1 & 1 & 0 & 0 \\ 0 & 1 & 0 & -1 & 1 & 0 \\ 2 & 1 & 2 & 0 & 0 & 1 \end{array}\right]$ $\left[\begin{array}{ccc|ccc} 1 & 0 & 1 & 1 & 0 & 0 \\ 0 & 1 & 0 & -1 & 1 & 0 \\ 0 & 1 & 0 & -2 & 0 & 1 \end{array}\right]$

$$\left[\begin{array}{ccc|ccc} 1 & 0 & 1 & 1 & 0 & 0 \\ 0 & 1 & 0 & -1 & 1 & 1 \\ 0 & 0 & 0 & -1 & -1 & 1 \end{array}\right]$$

Since we can never transform what is to the left of | into I_3,
A^{-1} does not exist.

Section 2.6

1. Expansion by row 2 yields

$(-1)^{2+1}(4)(-6) + (-1)^{2+2}(5)(-12) + (-1)^{2+3}6(-6) = 0$
Expansion by row 3 yields

$$(-1)^{3+1}(7)(-3) + (-1)^{3+2}(8)(-6) + (-1)^{3+3}(9)(-3) = 0$$

Chapter 3 Solutions

Section 3.1

1. max $z = 30x_1 + 100x_2$

   s.t. $x_1 + x_2 \leq 7$ (Land Constraint)

   $4x_1 + 10x_2 \leq 40$(Labor Constraint)

   $10x_1 \geq 30$(Govt. Constraint)

   $x_1 \geq 0,\ x_2 \geq 0$

3. 1 bushel of corn uses 1/10 acre of land and 4/10 hours of labor while 1 bushel of wheat uses 1/25 acre of land and 10/25 hours of labor. This yields the following formulation:

max $z = 3x_1 + 4x_2$

s.t. $x_1/10 + x_2/25 \leq 7$ (Land Constraint)

$4x_1/10 + 10x_2/25 \leq 40$ (Labor Const.)

$x_1 \geq 30$ (Govt. Const.)

$x_1 \geq 0,\ x_2 \geq 0$

Section 3.2

1. EF is $4x_1 + 10x_2 = 40$, CD is $x_1 = 3$, and AB is $x_1 + x_2 = 7$. The feasible region is bounded by ACGH. The dotted line in graph is isoprofit line $30x_1 + 100x_2 = 120$. Point G is optimal. At G the constraints $10x_1 \geq 30$ and $4x_1 + 10x_2 \leq 40$ are binding. Thus optimal solution has $x_1 = 3$, $x_2 = 2.8$ and $z = 30(3) + 100(2.8) = 370$.

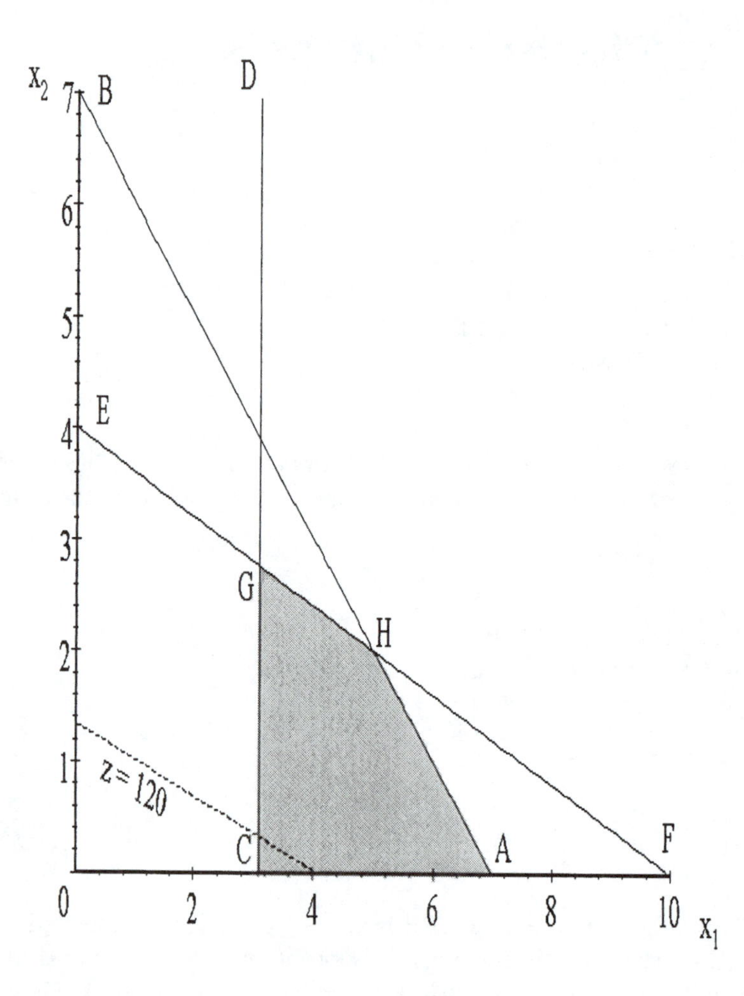

3. x_1 = Number of hours of Process 1 and x_2 = Number of hours of Process 2. Then the appropriate LP is

min $z = 4x_1 + x_2$

s.t. $3x_1 + x_2 \geq 10$ (A constraint)

$x_1 + x_2 \geq 5$ (B constraint)

$x_1 \geq 3$ (C constraint)

$x_1\ x_2 \geq 0$

AB is $3x_1 + x_2 = 10$. CD is $x_1 + x_2 = 5$. EF is $x_1 = 3$. The feasible region is shaded. Dotted line is isocost line $4x_1 + x_2 = 24$. Moving isocost line down to left we see that H (where B and C constraints intersect) is optimal. Thus optimal solution to LP is $x_1 = 3$, $x_2 = 2$, $z = 4(3) + 2 = \$14$.

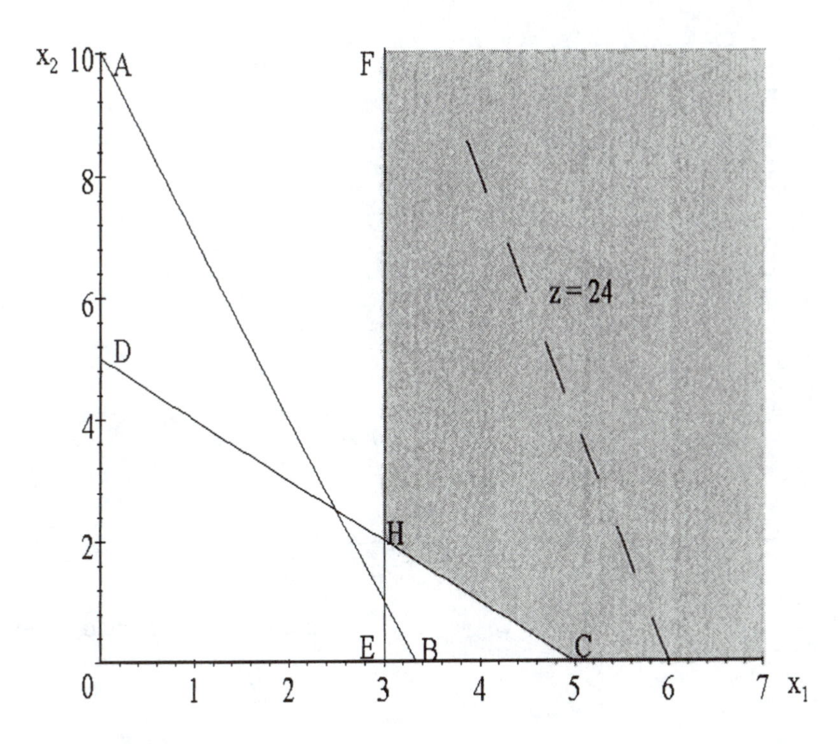

Section 3.3

1. AB is $x_1 + x_2 = 4$. CD is $x_1 - x_2 = 5$. From graph we see that there is no feasible solution (Case 3).

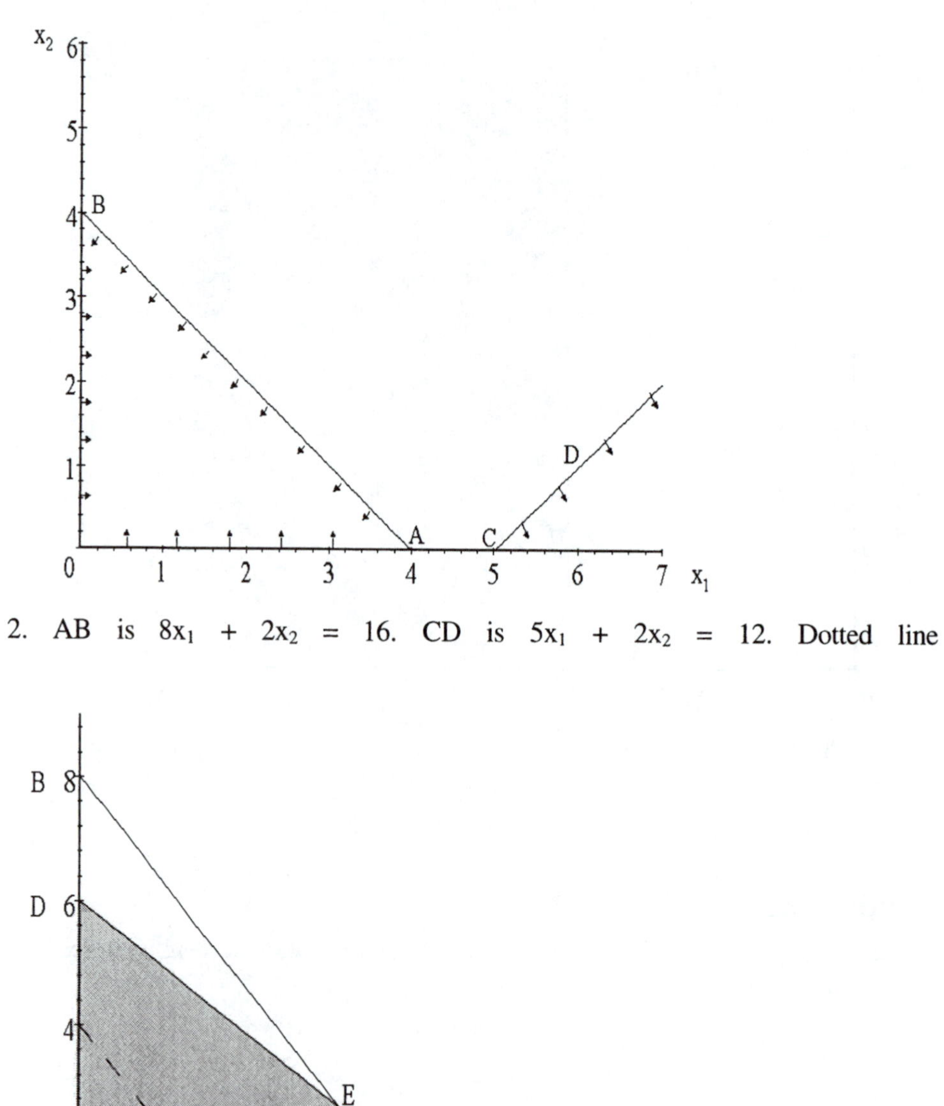

2. AB is $8x_1 + 2x_2 = 16$. CD is $5x_1 + 2x_2 = 12$. Dotted line

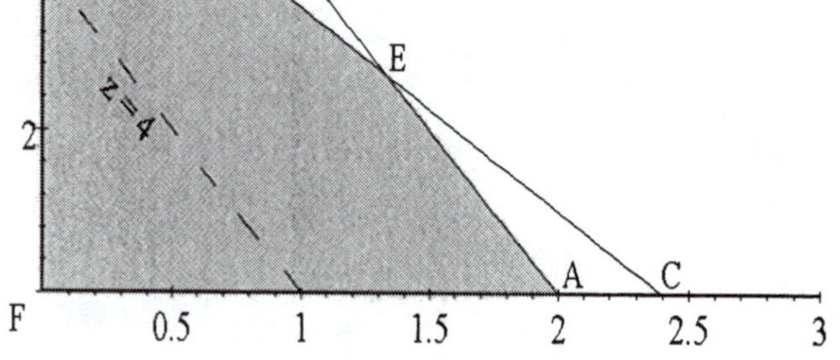

is $z = 4x_1 + x_2 = 4$. Feasible region is bounded by AEDF. Since isoprofit line is parallel to AE, entire line segment AE is optimal. Thus we have alternative or multiple optimal solutions.

3. AB is $x_1 - x_2 = 4$. AC is $x_1 + 2x_2 = 4$. Feasible region is bounded by AC and infinite line segment AB. Dotted line is isoprofit line $z = 0$. To increase z we move parallel to isoprofit line in an upward and `leftward' direction. We will never entirely lose contact with the feasible region, so we have an unbounded LP (Case 4).

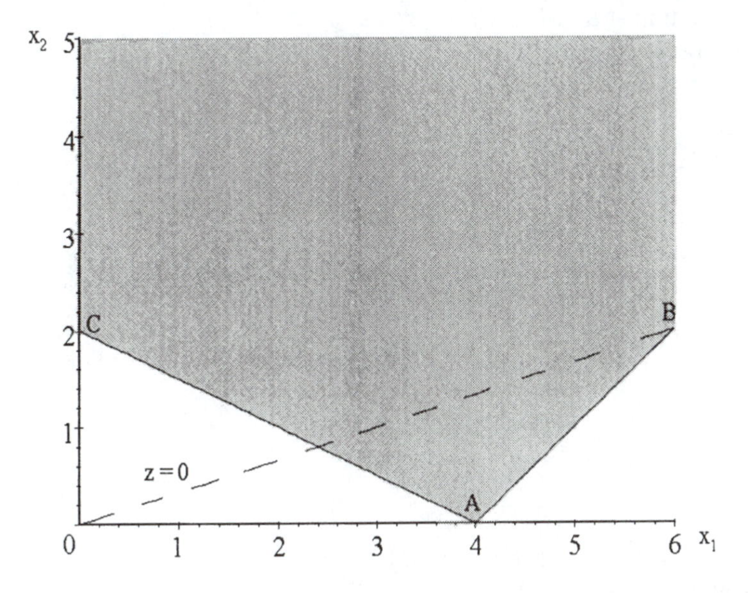

Section 3.4

1. For i = 1, 2, 3 let x_i = Tons of processed Factory i waste.
Then appropriate LP is

$$\min z = 15x_1 + 10x_2 + 20x_3$$
$$\text{s.t.} \quad .10x_1 + .20x_2 + .40x_3 \geq 30 \text{(Pollutant 1)}$$
$$.45x_1 + .25x_2 + .30x_3 \geq 40 \text{(Pollutant 2)}$$
$$x_1 \geq 0, x_2 \geq 0, x_3 \geq 0$$

It is doubtful that the processing cost is proportional to the amount of waste processed. For example, processing 10 tons of waste is probably not ten times as costly as processing 1 ton of waste. The divisibility and certainty assumptions seem reasonable.

Section 3.5

2. Let x_1 = employees starting at midnight

x_2 = employees starting at 4 AM

x_3 = employees starting at 8 AM

x_4 = employees starting at noon

x_5 = employees starting at 4 PM

x_6 = employees starting at 8 PM

Then a correct formulation is

$$
\begin{aligned}
\min z = \; & x_1 + x_2 + x_3 + x_4 + x_5 + x_6 \\
\text{s.t.} \; & x_1 + x_6 \geq 8 \\
& x_1 + x_2 \geq 7 \\
& x_2 + x_3 \geq 6 \\
& x_3 + x_4 \geq 6 \\
& x_4 + x_5 \geq 5 \\
& x_5 + x_6 \geq 4 \\
& x_1, x_2, x_3, x_4, x_5, x_6 \geq 0
\end{aligned}
$$

Section 3.6

2. NPV of Investment 1 = $-6 - 5/1.1 + 7/(1.1)^2 + 9/(1.1)^3 =$ \$2.00

NPV of Investment 2 = $-8 - 3/(1.1) + 9/(1.1)^2 + 7/(1.1)^3 =$ \$1.97

Let x_1 = Fraction of Investment 1 that is undertaken and

x_2 = Fraction of Investment 2 that is undertaken. If we measure NPV in thousands of dollars, we wish to solve the following LP:

$$
\begin{aligned}
\max z = \; & 2x_1 + 1.97x_2 \\
\text{s.t.} \; & 6x_1 + 8x_2 \leq 10 \\
& 5x_1 + 3x_2 \leq 7 \\
& x_1 \leq 1 \\
& x_2 \leq 1
\end{aligned}
$$

From the following graph we find the optimal solution to this LP to be $x_1 = 1$, $x_2 = .5$, $z =$ \$2,985.

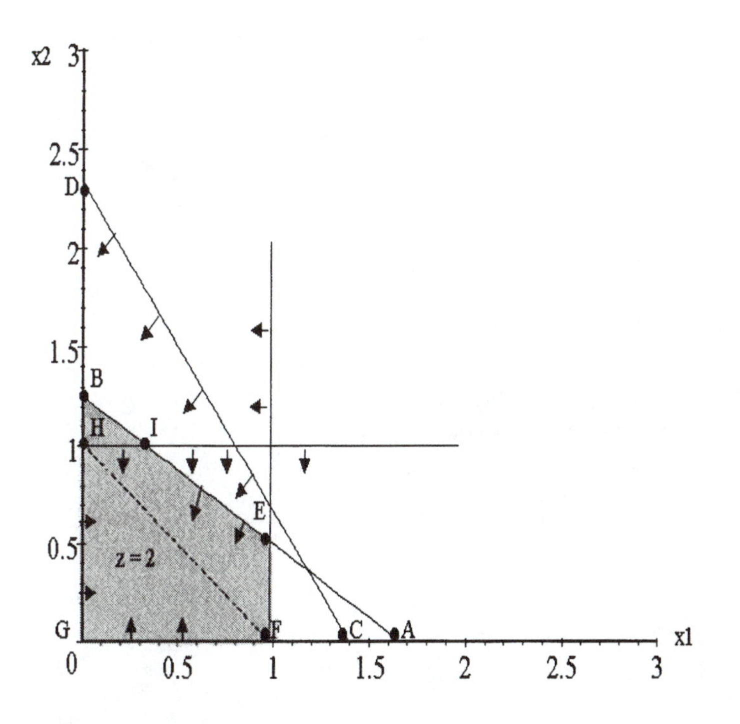

Section 3.8

1. Let (all variables are in ounces) Ing. 1 = Sugar, Ing. 2 = Nuts, Ing. 3 = Chocolate, Candy 1 = Slugger and Candy 2 = Easy Out Let x_{ij} = Ounces of Ing. i used to make candy j.

The appropriate LP is

max $z = 25(x_{12} + x_{22} + x_{32}) + 20(x_{11} + x_{21} + x_{31})$

s.t. $x_{11} + x_{12} \leq 100$ (Sugar Const.)

$x_{21} + x_{22} \leq 20$ (Nuts Constraint)

$x_{31} + x_{32} \leq 30$ (Chocolate Const.)

(1) $x_{22}/(x_{12} + x_{22} + x_{32}) \geq .20$

(2) $x_{21}/(x_{11} + x_{21} + x_{31}) \geq .10$

(3) $x_{31}/(x_{11} + x_{21} + x_{31}) \geq .10$

(1) -(3) are not LP constraints and should be replaced by the following three constraints:

Replace (1) by $x_{22} \geq .2(x_{12}+x_{22}+x_{32})$ or $.8x_{22}-.2x_{12}-.2x_{32} \geq 0$

Replace (2) by $x_{21} \geq .1(x_{11}+x_{21}+x_{31})$ or $.9x_{21}-.1x_{11}-.1x_{31} \geq 0$

Replace (3) by $x_{31} \geq .1(x_{11}+x_{21}+x_{31})$ or $.9x_{31}-.1x_{11}-.1x_{21} \geq 0$

5. xij = barrels of oil i used to make product j (j = 1 is gasoline and j = 2 is heating oil)

ai = dollars spent advertising product i

max z = 25(x11 + x21) + 20(x12 + x22) - a1 - a2

s.t. x11 + x21 = 5a1, x12 + x22 = 10a2, 2x11 - 3x21≥0 x11 + x12≤5000 x21 + x22≤10,000
4x12 - x22≥0 All variables ≥0.

Section 3.9

1. Let x_1 = Hours of Process 1 run per week.
   x_2 = Hours of Process 2 run per week.
   x_3 = Hours of Process 3 run per week.
   g_2 = Barrels of Gas 2 sold per week.
   o_1 = Barrels of Oil 1 purchased per week
   o_2 = Barrels of Oil 2 purchased per week

max $z = 9(2x_1)+10g_2+24(2x_3)-5x_1-4x_2-x_3-2o_1-3o_2$
s.t. $o_1 = 2x_1 + x_2$
$o_2 = 3x_1+3x_2+2x_3$
$o_1 \leq 200$
$o_2 \leq 300$
$g_2+3x_3 = x_1+3x_2$(Gas 2 Prod.)
$x_1+x_2+x_3 \leq 100$ (100 hours per wk. of cracker time)
All variables ≥0.

3. Let x6 = pounds of raw material used to produce Brute
   x7 = pounds of raw material used to produce Chanelle

Then the appropriate formulation is:

max z = 7x1 + 14x2 + 6x3 + 10x4 - 3x5
s.t. x5 ≤ 4000
3x2 + 2x4 + x5 ≤ 6000
x1 + x2 = 3x6
x3 + x4 = 4x7
x5 = x6 + x7
All variables ≥0

Section 3.10

1. Let x_t = Production during month t and i_t = Inventory at end of month t.

min z $= 5x_1+8x_2+4x_3+7x_4+2i_1+2i_2+2i_3+2i_4-6i_4$
s.t. $i_1 = x_1 - 50$
$i_2 = i_1 + x_2 - 65$
$i_3 = i_2 + x_3 - 100$
$i_4 = i_3 + x_4 - 70$
All variables ≥0.

3. Let Ct = cheesecakes baked during month t

Bt = black forest cakes baked during month t

It = number of cheesecakes in inventory at end of month t

It'= number of black forest cakes in inventory at the end of month t.

Then an appropriate formulation is

$$\min z = 3C1 + 3.4C2 + 3.8C3 + 2.5B1 + 2.8B2 + 3.4B3 + .5(I1 + I2 + I3) + .4(I1' + I2' + I3')$$

s.t.

$C1 + B1 \leq 65$

$C2 + B2 \leq 65$

$C3 + B3 \leq 65$

$I1 = C1 - 40$

$I2 = I1 + C2 - 30$

$I3 = I2 + C3 - 20$

$I1' = B1 - 20$

$I2' = I1' + B2 - 30$

$I3' = I2' + B3 - 10$

All variables ≥ 0

Section 3.11

3. Let A = \$ invested in A.

B = \$ invested in B.

c_0 = Leftover cash at Time 0

c_1 = Leftover cash at Time 1

c_2 = Leftover cash at Time 2.

Then a correct formulation is

$\max z = c_2 + 1.9B$

s.t. $A + c_0 = 10{,}000$ (Time 0 Avail = Time 0 Inv.)

$.2A + c_0 = B + c_1$ (Time 1 Avail. = Time 1 Inv.)

$1.5A + c_1 = c_2$ (Time 2 Avail. = Time 2 Inv.)

All Variables ≥ 0.

The optimal solution to this LP is $B = c_0 = \$10{,}000$, $A = c_1 = c_2 = 0$ and $z = \$19{,}000$. Notice that it is optimal to `wait' for the `good' investment (B) even though leftover cash earns no interest.

Section 3.12

2. Let JAN1 = Number of computers rented at beginning of JAN for one month, etc. Also define IJAN = Number of computers available to meet January demand, etc. The appropriate LP is

min z = 200(JAN1 + FEB1 + MAR1 + APR1 + MAY1 + JUNE1) + 350(JAN2 + FEB2 + MAR2 + APR2
+MAY2+JUN2)+450(JAN3+FEB3+MAR3+APR3+MAY3+JUN3)-150MAY3-30
0JUN3
-175JUN2

s.t. IJAN = JAN1 + JAN2 + JAN3
IFEB = IJAN - JAN1 + FEB1 + FEB2 + FEB3
IMAR = IFEB - JAN2 - FEB1 + MAR1 + MAR2 + MAR3
IAPR = IMAR - FEB2 - MAR1 - JAN3 + APR1 + APR2 + APR3
IMAY = IAPR - FEB3 - MAR2 - APR1 + MAY1 + MAY2 + MAY3
IJUN = IMAY - MAR3 - APR2 - MAY1 + JUN1 + JUN2 + JUN3
IJAN≥9
IFEB≥5
IMAR≥7
IAPR≥9
IMAY≥10
IJUN≥5 All variables ≥0

Chapter 4 Solutions

Section 4.1

1. $\max z = 3x_1 + 2x_2$

s.t. $2x_1 + x_2 + s_1 = 100$

$x_1 + x_2 + s_2 = 80$

$x_1 + s_3 = 40$

Section 4.4

1. From Figure 2 of Chapter 3 we see that the extreme points of the feasible region are

Basic Feasible Solution

$H = (0, 0)$ $s_1 = 100, s_2 = 80, s_3 = 40$ $x_1 = x_2 = x_3 = 0$
$E = (40, 0)$ $x_1 = 40, s_1 = 20, s_2 = 40$ $x_2 = x_3 = s_3 = 0$
$F = (40, 20)$ $x_1 = 40, x_2 = 20, s_2 = 20$ $x_3 = s_1 = s_3 = 0$
$G = (20, 60)$ $x_1 = 20, x_2 = 60, s_3 = 20$ $x_3 = s_1 = s_2 = 0$
$D = (0, 80)$ $s_1 = 20, x_2 = 80, s_3 = 40$ $s_2 = x_1 = x_3 = 0$

Section 4.5

2.

z	x_1	x_2	s_1	s_2	RHS	Ratio
1	-2	-3	0	0	0	
0	1	2	1	0	6	3* Enter x_2 in row 1
0	2	1	0	1	8	8
0	-1/2	0	3/2	0	9	
0	1/2	1	1/2	0	3	6
0	3/2	0	-1/2	1	5	10/3* Enter x_1 in row 2

z	x_1	x_2	s_1	s_2	RHS	Ratio
0	0	0	4/3	1/3	32/3	

0   0   1 2/3 -1/3   4/3

---

0   1   0 -1/3 2/3 10/3

This is an optimal tableau and the optimal solution is z = 32/3, x_1 = 10/3, x_2 = 4/3, $s_1 = s_2 = 0$.

Section 4.6

3.

Z	X1	X2	S1	S2	RHS
1	-2	5	0	0	0
0	3	8	1	0	12
0	2	3	0	1	6

We enter X2 into the basis in Row 1. The resulting optimal tableau is

Z	X1	X2	S1	S2	RHS
1	-31/8	0	-5/8	0	-7.5
0	3/8	1	1/8	0	1.5
0	7/8	-7/3	-3/8	1	1.5

The LP's optimal solution is Z= -7.5 , X1=0, and X2=1.5.

Section 4.7

5.

Z	X1	X2	S1	S2	RHS
1	-2	-2	0	0	0
0	1	1	1	0	6
0	2	1	0	1	13

We now arbitrarily choose to enter X1 into basis. X1 enters basis in Row 1 yielding following optimal tableau.

Z	X1	X2	S1	S2	RHS
1	0	0	2	0	12
0	1	1	1	0	6
0	0	-1	-2	1	1

This tableau yields the optimal solution Z = 12, X1 = 6, X2=0. Pivoting X2 into the basis yields the alternative optimal solution Z=12, X1=0, X2=6. All optimal solutions are of the form c(1st optimal solution) + (1-c)(2nd optimal solution) where 0<=c<=1. This shows all optimal solutions are of form Z=12, X1= 6c, X2=6-6c, 0<=c<=1.

Section 4.8

6.

Z	X1	X2	S1	S2	RHS
1	1	3	0	0	0
0	1	-2	1	0	4
0	-1	1	0	1	3

X2 enters the basis in ROW 2 yielding the following tableau:

Z	X1	X2	S1	S2	RHS
1	4	0	0	-3	-9
0	-1	0	1	2	10
0	-1	1	0	1	3

We would like to enter X1 into the basis but there is no row in which X1 has a positive coefficient. Therefore the LP is unbounded.

Section 4.10

4. MODEL:

```
SETS:
PRODUCTS/1..3/:MADE,PROFIT;
RESOURCES/1..3/:AVAIL;
RESPRO(RESOURCES,PRODUCTS):USAGE;
ENDSETS
MAX=@SUM(PRODUCTS(I):PROFIT(I)*MADE(I));
@FOR(RESOURCES(I):@SUM(PRODUCTS(J):MADE(J)*USAGE(I,J))<=AV
AIL(I));
DATA:
PROFIT= 800,1500,2500;
AVAIL=50,10 150;
USAGE= 2,3,5
    .3,.7,.2
     120,12,20;
ENDDATA
END
```

Optimal Solution is
Global optimal solution found at step: 1
Objective value: 18750.00

Variable	Value	Reduced Cost
MADE(1)	0.0000000	14200.00
MADE(2)	0.0000000	0.0000000
MADE(3)	7.500000	0.0000000

Section 4.11

4. Here are the pivots:

Z	X1	X2	X3	X4	S1	S2	S3	RHS
1	3	-1	6	0	0	0	0	0
0	9	1	-9	-2	1	0	0	0
0	1	1/3	-2	-1/3	0	1	0	0
0	-9	-1	9	2	0	0	1	1

X2 now enters in Row 1 yielding the following tableau.

Z	X1	X2	X3	X4	S1	S2	S3	RHS
1	12	0	-3	-2	1	0	0	0
0	9	1	-9	-2	1	0	0	0
0	-2	0	1	1/3	-1/3	1	0	0
0	0	0	0	0	1	0	1	1

We now enter X3 into the basis in Row 2.

Z	X1	X2	X3	X4	S1	S2	S3	RHS
1	6	0	0	-1	0	3	0	0
0	-9	1	0	1	-2	9	0	0
0	-2	0	1	1/3	-1/3	1	0	0
0	0	0	0	0	1	0	1	1

We now enter X4 into the basis and arbitrarily choose to enter X4 in Row 1.

Z	X1	X2	X3	X4	S1	S2	S3	RHS
1	-3	1	0	0	-2	12	0	0
0	-9	1	0	1	-2	9	0	0
0	1	-1/3	1	0	1/3	-2	0	0
0	0	0	0	0	1	0	1	1

X1 now enters basis in Row 2.

Z	X1	X2	X3	X4	S1	S2	S3	RHS
1	0	0	3	0	-1	6	0	0
0	0	-2	9	1	1	-9	0	0
0	1	-1/3	1	0	1/3	-2	0	0
0	0	0	0	0	1	0	1	1

We now choose to enter S1 in Row 1.

Z	X1	X2	X3	X4	S1	S2	S3	RHS
1	0	-2	12	1	0	-3	0	0
0	0	-2	9	1	1	-9	0	0
0	1	1/3	-2	-1/3	0	1	0	0
0	0	2	-9	-1	0	9	1	1

S2 would now enter basis in Row 2. This will bring us back to initial tableau, so cycling has occurred.

Section 4.12

5. Initial tableau is

Z	X1	X2	X3	A1	A2	RHS
1	3M-1	2M-1	3M	0	0	6M
0	2	1	1	1	0	4
0	1	1	2	0	1	2

We now enter X3 in the basis in row 2:

Z	X1	X2	X3	A1	A2	RHS
1	3M/2-1	M/2-1	0	0	-3M/2	3M
0	3/2	1/2	0	1	-1/2	3
0	1/2	1/2	1	0	1/2	1

We now enter X1 in the basis in row 2.

Z	X1	X2	X3	A1	A2	RHS
1	0	-M	2-3M	0	1-3M	2
0	0	-1	-3	1	-2	0
0	1	1	2	0	1	2

This optimal tableau yields the optimal solution Z=2, X1=2, X2=X3=0.

Section 4.13

5. Initial Phase I Tableau is

W	X1	X2	X3	A1	A2	RHS
1	3	2	3	0	0	6
0	2	1	1	1	0	4
0	1	1	2	0	1	2

X1 enters the basis in Row 1 yielding

W	X1	X2	X3	A1	A2	RHS
1	0	1/2	3/2	-3/2	0	0
0	1	1/2	1/2	1/2	0	2
0	0	1/2	3/2	-1/2	1	0

This completes Phase I. This is Case III so we now drop the A1 column and begin Phase II with A2 as a basic variable.

Z	X1	X2	X3	A2	RHS
1	0	-1/2	1/2	0	2
0	1	1/2	1/2	0	2
0	0	1/2	3/2	1	0

X2 now enters basis in Row 2 yielding the following optimal tableau:

Z	X1	X2	X3	A2	RHS
1	0	0	2	1	2
0	1	0	-1	-1	2
0	0	1	3	2	0

The optimal solution is Z=2, X1=2, X2=X3=0.

Section 4.14

1. Let $i_t = i_t' - i_t''$ be the inventory position at the end of month t. For each constraint in original problem replace i_t by $i_t' - i_t''$. Also add the sign restrictions $i_t' >= 0$ and $i_t'' >= 0$. To ensure that demand is met by end of Quarter 4 add sign restriction $i_4' - i_4'' >= 0$. Change the terms involving i_t in objective function to $100i_1' + 110i_1'' + 100i_2' + 110i_2'' + 100i_3' + 110i_3'' + 100i_4' + 110i_4''$.

Section 4.16

1a. The only point satisfying the LIP constraint and the budget constraint is (6, 0). Thus $x_1 = 6$, $x_2 = 0$ is the optimal solution. This leaves the HIW goal unmet by 5 million.

1b. Only point satisfying the HIM constraint, the budget line, and the LIP constraint, is the point (6,0). Thus

$x_1 = 6$, $x_2 = 0$ is the optimal solution.

1c. The point satisfying both the budget constraint and the HIM(highest priority goal)goal that is closest to the HIW constraint occurs where the HIM and budget lines intersect. This is at the point $x_1 = 5$ and $x_2 = 5/3$.

1d. The point satisfying the budget constraint and HIW that is closest to HIM is C. Thus $x_1 = 3$ and $x_2 = 5$ is the optimal solution.

5. Let H = pounds of head in mixture
C = pounds of chuck in mixture
MU = pounds of mutton in mixture
MO = pounds of moisture in mixture
Then the appropriate goal programming formulation is

$\min P_1s_1^- + P_2s_2^+ + P_3s_3^+$

st. $2H + .26C + .08MU + s_1^- - s_1^+ = 15$ (protein)
$.05H + .24C + .11MU + s_2^- - s_2^+ = 8$ (fat)
$12H + 9C + 8MU + s_3^- - s_3^+ = 800$ (cost)
$H + C + MU + MO = 100$
All variables nonnegative

Section 4.17

1. See file S4_17_1.xls

	A	B	C	D	E	F	G
1	SECTION 4-17						
2	PROBLEM 1						
3		SUPP1	SUPP2	SUPP3	TOTALCOST		
4	AMOUNT	1200	0	100	6300		
5	COST	5	4	3	BOUGHT		NEEDED
6	LARGE	0.4	0.3	0.2	500	=>=	500
7	MEDIUM	0.4	0.35	0.2	500	>=	300
8	SMALL	0.2	0.35	0.6	300	=>=	300

2.

	A	B	C	D	E	F	G	H	I
1	PROBLEM 2								
2	SECTION 4-17								
3									
4									
5	START		MON	TUES	WED	THUR	FRI	SAT	SUN
6	MON	6.3333333	1	1	1	1	1	0	0
7	TUES	5	0	1	1	1	1	1	0
8	WED	0.3333333	0	0	1	1	1	1	1
9	THURS	7.3333333	1	0	0	1	1	1	1
10	FRI	0	1	1	0	0	1	1	1
11	SAT	3.3333333	1	1	1	0	0	1	1
12	SUN	0	1	1	1	1	0	0	1
13	TOTAL	22.333333							
14									
15		AVAILABLE	17	14.667	15	19	19	16	11
16			=>=	>=	=>=	=>=	>=	=>=	=>=
17		NEEDED	17	13	15	19	14	16	11

See file S4_17_2.xls

Chapter 5 Solutions

Section 5.1

1. Typical isoprofit line is $3x_1+c_2x_2=z$. This has slope $-3/c_2$. If slope of isoprofit line is <-2, then Point C is optimal. Thus if $-3/c_2<-2$ or $c_2<1.5$ the current basis is no longer optimal. Also if the slope of the isoprofit line is >-1 Point A will be optimal. Thus if $-3/c_2>-1$ or $c_2>3$ the current basis is no longer optimal. Thus for $1.5\leq c_2\leq 3$ the current basis remains optimal.
For $c_2 = 2.5$ $x_1 = 20$, $x_2 = 60$, but $z = 3(20) + 2.5(60) = \$210$.

2. Currently Number of Available Carpentry Hours = $b_2 = 80$. If we reduce the number of available carpentry hours we see that when the carpentry constraint moves past the point (40, 20) the carpentry and finishing hours constraints will be binding at a point where $x_1>40$. In this situation $b_2<40 + 20 = 60$. Thus for $b_2<60$ the current basis is no longer optimal. If we increase the number of available carpentry hours we see that when the carpentry constraint moves past (0, 100) the carpentry and finishing hours constraints will both be binding at a point where $x_1<0$. In this situation $b_2>100$. Thus if $b_2>100$ the current basis is no longer optimal. Thus the current basis remains optimal for $60\leq b_2\leq 100$. If $60\leq b_2\leq 100$, the number of soldiers and trains produced will change.

Section 5.2

1a. 4250 - 5(75) = \$3875. Note we are in allowable range because land can decrease by up to 6.667 acres.

1b. Optimal solution is still to plant 25 acres of wheat and 20 acres of corn and use 350 labor hours.
New Profit = 130(25) + 200(20) - 10(350) = \$3750 or New Profit=Old Profit -20 (25) = \$3750 (This follows because profit coefficient for AI is decreased by 20 and AD=30)

1c. Allowable Decrease = SLACK =15 and Allowable Increase = Infinity. 130 bushels is in allowable range so solution remains unchanged.

2a. Decision variables remain the same.
New z-Value = Old z-value+10(88) = \$33,420

2b. Relevant Shadow Price is -\$20. Current basis remains optimal if demand is decreased by up to 3 cars, so Dual Price may be used to compute new z-value.
New Profit = \$32,540 + (-2) (-20) = \$33,580.

Section 5.3

2. Cannot answer this question because current basis is no longer optimal if one more machine is available.

3. If you were given an extra ounce of chocolate, this would reduce cost by 2.5 cents, so you would be willing to pay up to 2.5 cents for an ounce of chocolate.

4. Plant 1 shadow price is \$2, so another unit of capacity at Plant 1 reduces cost by \$2. Thus company can pay up to \$2 for an extra unit of capacity at Plant 1 and still be better off.

Section 5.4

3.

	Optimal Solution	Optimal z-value
$0 \le c_2 \le 1.5$	$x_1 = 40$ $x_2 = 20$	$120 + 20c_2$
$1.5 \le c_2 \le 3$	$x_1 = 20$ $x_2 = 60$	$60 + 60c_2$
$c_2 \ge 3$	$x_1 = 0$ $x_2 = 80$	$80c_2$

Also see graphs on the following pages.

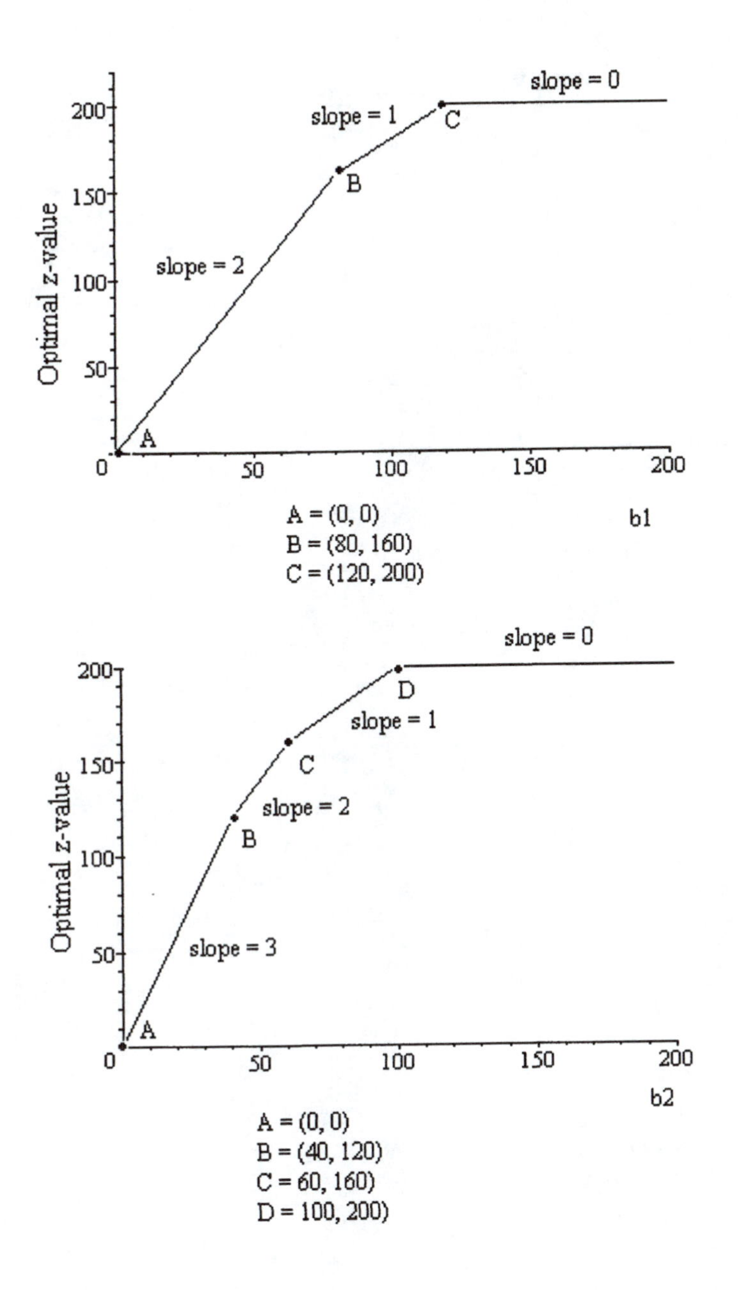
slope = 0
200
slope = 1
C
150
B
Optimal z-value
100
slope = 2
50
A
0
50
100
150
200
A = (0, 0)
b1
B = (80, 160)
C = (120, 200)
slope = 0
200
D
slope = 1
150
C
slope = 2
Optimal z-value
100
B
50
slope = 3
A
0
50
100
150
200
b2
A = (0, 0)
B = (40, 120)
C = 60, 160)
D = 100, 200)

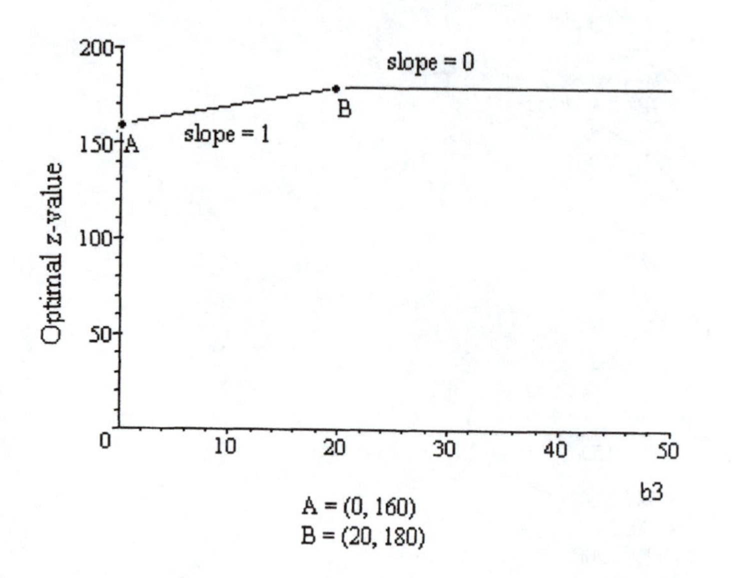

200
150
100
50
0
Optimal z-value
slope = 0
slope = 1
A
B
10
20
30
40
50
b3
A = (0, 160)
B = (20, 180)

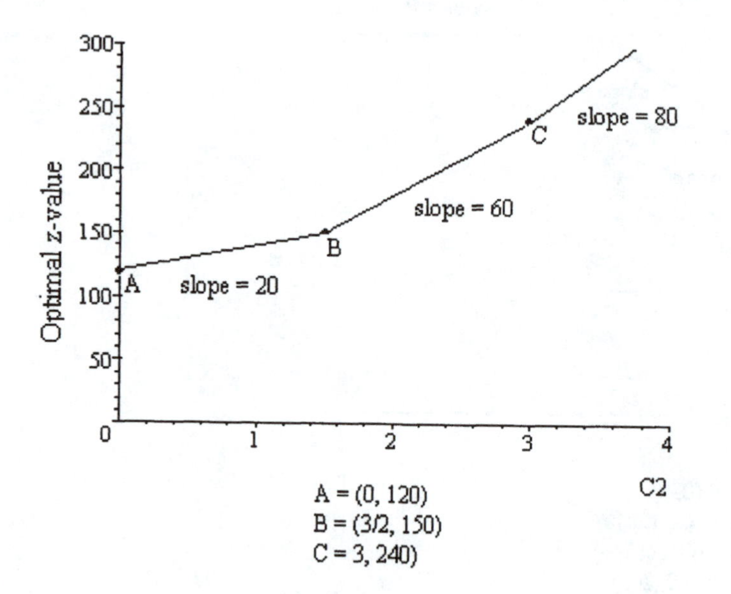

300
250
200
150
100
50
0
Optimal z-value
slope = 80
slope = 60
slope = 20
A
B
C
1
2
3
4
C2
A = (0, 120)
B = (3/2, 150)
C = 3, 240)

Chapter 6 Solutions

Section 6.1

1. Typical isoprofit line is $3x_1+c_2x_2=z$. This has slope $-3/c_2$. If slope of isoprofit line is <-2, then Point C is optimal. Thus if $-3/c_2<-2$ or $c_2<1.5$ the current basis is no longer optimal. Also if the slope of the isoprofit line is >-1 Point A will be optimal. Thus if $-3/c_2>-1$ or $c_2>3$ the current basis is no longer optimal. Thus for $1.5\le c_2\le 3$ the current basis remains optimal.
For $c_2 = 2.5$ $x_1 = 20$, $x_2 = 60$, but $z = 3(20) + 2.5(60) = \$210$.

2. Currently Number of Available Carpentry Hours = $b_2 = 80$. If we reduce the number of available carpentry hours we see that when the carpentry constraint moves past the point (40, 20) the carpentry and finishing hours constraints will be binding at a point where $x_1>40$. In this situation $b_2<40 + 20 = 60$. Thus for $b_2<60$ the current basis is no longer optimal. If we increase the number of available carpentry hours we see that when the carpentry constraint moves past (0, 100) the carpentry and finishing hours constraints will both be binding at a point where $x_1<0$. In this situation $b_2>100$. Thus if $b_2>100$ the current basis is no longer optimal. Thus the current basis remains optimal for $60\le b_2\le 100$. If $60\le b_2\le 100$, the number of soldiers and trains produced will change.

Section 6.2

1. BV = $\{x_1, x_2\}$ $B = \begin{bmatrix} 2 & -1 \\ -1 & 1 \end{bmatrix}$ $B^{-1} = \begin{bmatrix} 1 & 1 \\ 1 & 2 \end{bmatrix}$ $c_{BV} = [3\ 1]$

$c_{BV}B^{-1} = [4\ 5]$
Coefficient of s_1 in row 0 = 4
Coefficient of s_2 in row 0 = 5

Right hand Side of row 0 = $\mathbf{c_{BV}}B^{-1}\mathbf{b} = [4\ 5]\begin{bmatrix} 2 \\ 4 \end{bmatrix} = 28$

s_1 column = $B^{-1}\begin{bmatrix} 1 \\ 0 \end{bmatrix} = \begin{bmatrix} 1 \\ 1 \end{bmatrix}$

s_2 column = $B^{-1}\begin{bmatrix}0\\1\end{bmatrix} = \begin{bmatrix}1\\2\end{bmatrix}$

x_1 column = $\begin{bmatrix}1\\0\end{bmatrix}$ x_2 column = $\begin{bmatrix}0\\1\end{bmatrix}$

Right hand Side of Constraints = $B^{-1}\mathbf{b} = \begin{bmatrix}1 & 1\\1 & 2\end{bmatrix}\begin{bmatrix}2\\4\end{bmatrix} = \begin{bmatrix}6\\10\end{bmatrix}$

Thus the optimal tableau is

$z + 4s_1 + 5s_2 = 28$
$x_1 + s_1 + s_2 = 6$
$x_2 + s_1 + 2s_2 = 10$

Section 6.3

6a. x_1 is non-basic so changing the coefficient of x_1 in the objective function will only change the coefficient of x_1 in the optimal row 0. Let the new coefficient of x_1 in the objective function be $3 + \Delta$. The new coefficient of x_1 in the optimal row 0 will be
$\mathbf{c}_{BV}B^{-1}\mathbf{a}_1-(3 + \Delta) = 3 - \Delta$. Thus if $3 - \Delta\geq 0$ or $\Delta\leq 3$ the current basis remains optimal. Thus if profit for a Type 1 Candy Bar is ≤ 6 cents the current basis remains optimal.

6b. Changing Candy Bar 2 profit to $7 + \Delta$ changes $\mathbf{c}_{BV}B^{-1}$ to

$[5\ \ 7 + \Delta]\begin{bmatrix}3/2 & -1/2\\-1/2 & 1/2\end{bmatrix} = [4 - \Delta/2\ \ 1 + \Delta/2]$

Then coefficient of x_1 in row 0 = $[4-\Delta/2\ \ 1 +\Delta/2]\begin{bmatrix}1\\-3\\2\end{bmatrix} = 3 + \Delta/2$.

Thus row 0 of optimal tableau is now
$z + (3 + \Delta/2)\,x_1 + (4- \Delta/2)\,s_1 + (1 + \Delta/2)s_2 =$?

Thus current basis remains optimal if (1)-(3) are met:
(1) $3 + \Delta/2\geq 0$ (or $\Delta\geq -6$)

(2) $4-\Delta/2 \geq 0$ (or $\Delta \leq 8$)
(3) $1+\Delta/2 \geq 0$ (or $\Delta \geq -2$)
Thus if $-2 \leq \Delta \leq 8$ the current basis remains optimal. Thus if profit for Type 2 Candy Bar is between 7-2 = 5 and 7 + 8 = 15 cents the current basis remains optimal.

6c. If the amount of sugar available is changed to $50+\Delta$ the current basis remains optimal iff

$$\begin{bmatrix} 3/2 & -1/2 \\ -1/2 & 1/2 \end{bmatrix} \begin{bmatrix} 50 + \Delta \\ 100 \end{bmatrix} = \begin{bmatrix} 25 + 3\Delta/2 \\ 25 - \Delta/2 \end{bmatrix}$$

Thus current basis remains optimal iff (1) - (2) hold
(1) $25+3\Delta/2 \geq 0$ (or $\Delta \geq -50/3$)
(2) $25-\Delta/2 \geq 0$ (or $\Delta \leq 50$.)
Thus current basis remains optimal iff $100/3 = 50-50/3 \leq$ Amount of Available Sugar $\leq 50+50 = 100$.

6d. After this change the current basis is still optimal.

$$\text{New Profit} = \mathbf{c}_{BV}B^{-1}\mathbf{b} = [4\ 1] \begin{bmatrix} 60 \\ 100 \end{bmatrix} = \$3.40$$

New values of decision variables are found from

$$\begin{bmatrix} x_3 \\ x_2 \end{bmatrix} = B^{-1}\mathbf{b} = \begin{bmatrix} 3/2 & -1/2 \\ -1/2 & 1/2 \end{bmatrix} \begin{bmatrix} 60 \\ 100 \end{bmatrix} = \begin{bmatrix} 40 \\ 20 \end{bmatrix}$$

Thus 40 Type 3 Candy Bars, 20 Type 2 Candy Bars, and 0 Type 1 candy bars would now be manufactured. If only 30 ounces of sugar were available the current basis would no longer be optimal and we would have to resolve the problem to find the new optimal solution.

6e. Coefficient of Type 1 Candy Bar in row 0 is now

$$[4\ 1] \begin{bmatrix} 1/2 \\ 1/2 \end{bmatrix} -3 = -.5.$$ Thus current basis is no longer optimal and the new optimal solution would manufacture Type 1 Candy Bars.

6f. The coefficient of Type 4 Candy Bars in row 0 will now be

$[4\ 1]\begin{bmatrix}3\\4\end{bmatrix} - 17 = -1$. Thus x_4 should be entered into the basis and the current basis is no longer optimal. The new optimal solution will make Type 4 Candy Bars.

Section 6.4

4. Both the fat and calorie constraints are nonbinding. The new fat and calorie constraints are each within their allowable range so the current basis remains optimal.

5. We need to use the 100% Rule. Since

$$\frac{30-15}{30} + \frac{80-60}{50} = .90 \le 1$$ the current basis remains optimal.

6. Use the 100% rule. Since

$$\frac{8-6}{4} + \frac{500-60}{\infty} = .50 \le 1$$ the current basis remains optimal.

Section 6.5

1.
$$\begin{aligned} \min w = {} & y_1 + 3y_2 + 4y_3 \\ \text{s.t.} \quad & -y_1 + y_2 + y_3 \ge 2 \\ & y_1 + y_2 - 2y_3 \ge 1 \\ & y_1, y_2, y_3 \ge 0 \end{aligned}$$

4.
$$\begin{aligned} \max z = {} & 6x_1 + 8x_2 \\ \text{s.t.} \quad & x_1 + x_2 \le 4 \\ & 2x_1 - x_2 \le 2 \\ & 2x_2 = -1 \\ & x_1 \le 0 \quad x_2 \text{ u.r.s.} \end{aligned}$$

Section 6.7

1a. min $w = 100y_1 + 80y_2 + 40y_3$

s.t. $2y_1 + y_2 + y_3 \geq 3$

$y_1 + y_2 \geq 2$

$y_1 \geq 0\ y_2 \geq 0\ y_3 \geq 0$

1b. and 1c. $y_1 = 1\ y_2 = 1\ y_3 = 0\ w = 180$. Observe that this solution has a w-value that equals the optimal primal z-value. Since this solution is dual feasible it must be optimal (by Lemma 2) for the dual.

4. The dual is

max $z = 4x_1 + 20x_2 + 10x_3$

s.t. $x_1/2 + x_2 + x_3 \geq 2$

$x_1/4 + 3x_2 + x_3 \geq 3$

$x_1, x_2, x_3 \geq 0$

The optimal solution to the dual is $x_1 = 0$, $x_2 = 1/2$, $x_3 = 3/2$, $z = 4(0) + 20(1/2) + 10(3/2) = 25$ cents.

Section 6.8

2a. Shadow Price for Sugar Const. = 4 Shadow Price for Chocolate Const. = 1. If 1 extra ounce of sugar were available profits would increase by $4. If 1 extra ounce of chocolate were available profits would increase by $1. Without further information however we cannot determine how much we would pay for an additional ounce of chocolate or sugar.

2b. From Section 6.3 current basis remains optimal if $100/3 \leq$ Available Sugar ≤ 100.
Here $\Delta b_1 = 60\text{-}50 = 10$ so
New z-value = [old z-value] + 10(4) = 340 cents

2c. $\Delta b_1 = -10$ Thus [New z-value] = 300-10 (4) = 260 cents.

2d. Since current basis is no longer optimal we cannot answer this question without resolving the problem.

Section 6.9

1. Dual constraint for computer tables is $6y_1 + 2y_2 + y_3 \geq 35$. Since [0 10 10] does not satisfy this constraint the current basis is no longer optimal. Another way of seeing it: a computer table uses \$20 worth of finishing time and \$10 worth of carpentry time. Since a computer table sells for \$35 it pays to make computer tables and the current basis is no longer optimal.

Section 6.10

1a. $\min w = 600y_1 + 400y_2 + 500y_3$

$$\begin{aligned} \text{s.t.}\quad & 4y_1 + y_2 + 3y_3 \geq 6 \ (1) \\ & 9y_1 + y_2 + 4y_3 \geq 10 \ (2) \\ & 7y_1 + 3y_2 + 2y_3 \geq 9 \ (3) \\ & 10y_1 + 40y_2 + y_3 \geq 20 \ (4) \\ & y_1, y_2, y_3, y_4 \geq 0 \end{aligned}$$

1b. Since $s_3>0$, $y_3 = 0$. Since $x_1>0$, (1) is binding. Since $x_4>0$, (4) is binding. Setting $y_3 = 0$ and solving (1) and (4) simultaneously yields $y_1 = 22/15$ $y_2 = 2/15$. Thus the optimal dual solution is $y_1 = 22/15$, $y_2 = 2/15$, $y_3 = 0$, $z = 2800/3$.

1c. (40): $s_3>0$ implies $y_3 = 0$. This is reasonable because if all available glass is not being used an additional ounce of glass will not increase z hence glass constraint should have 0 shadow price.

(41): $y_2>0$ implies $s_2 = 0$. Since $y_2>0$, an additional minute of packaging time will increase z. Thus all packaging time that is currently available must be used (hence $s_2 = 0$).

(42):$e_2 = 9(22/15) + 2/15 - 10 = 50/15>0$ implies $x_2 = 0$. Note that e_2 = [Cost of Resources Used to Make a Beer glass] -[Price of a Beer Glass]. Since $e_2>0$, producing a beer glass would not be profitable so $x_2 = 0$ should hold.

(43):$x_1>0$ implies $e_1 = 0$. Since $x_1>0$, we are manufacturing wine glasses. Thus MR = MC yields that Sale Price of a Wine Glass = Cost of Making a Wine Glass or [Cost of Making a Wine Glass]-[Sales Price of Wine Glass] = 0 Thus $e_1 = 4(22/15) + 1(2/15) + 3(0)-6 = 0$.

Section 6.11

3. The rhs of the optimal tableau is now

$$\begin{bmatrix} 1 & 2 & -8 \\ 0 & 2 & -4 \\ 0 & -.5 & 1.5 \end{bmatrix} \begin{bmatrix} 20 \\ 20 \\ 8 \end{bmatrix} = \begin{bmatrix} -4 \\ 8 \\ 2 \end{bmatrix}$$

rhs of row 0 is $[0\ 10\ 10] \begin{bmatrix} 20 \\ 20 \\ 8 \end{bmatrix} = 280$

We can apply the dual simplex to the following tableau:

z	x_1	x_2	x_3	s_1	s_2	s_3	RHS
1	0	5	0	0	10	10	280
0	0	-2	0	1	2	-8	-4
0	0	-2	1	0	2	-4	8
0	1	1.25	0	0	-.5	1.5	2
Ratio		2.5				1.25	

s_3 now enters the basis in row 1. This pivot yields an optimal tableau with $z = 275$ $s_3 = 1/2$ $x_3 = 10$ $x_1 = 1.25$.

Section 6.12

1. From printouts we find that only School 4 is inefficient. We find that .065(School 2 Inputs) + 1.21(School 3 Inputs) =

$$\begin{bmatrix} 14.22 \\ 7.59 \\ .08 \end{bmatrix}$$

Thus composite school uses 5% less of Inputs 1 and 2.

$$.065(\text{School 2 Outputs}) + 1.21(\text{School 3 Outputs}) = \begin{bmatrix} 13.96 \\ 9 \\ 10.13 \end{bmatrix}$$

Thus composite school can produce 55% better reading with fewer inputs.

Problem 1 School 4 Printout

```
MAX 9T1+9T2+9T3
ST
-9T1-7T2-6T3+13W1+4W2+.05W3>0
-10T1-8T2-7T3+14W1+5W2+.05W3>0
-11T1-7T2-8T3+11W1+6W2+.06W3>0
-9T1-9T2-9T3+15W1+8W2+.08W3>0
T1>.0001
T2>.0001
T3>.0001
W1>.0001
W2>.0001
W3>.0001
15W1+8W2+.08W3=1
END
```

```
LP OPTIMUM FOUND AT STEP      6

     OBJECTIVE FUNCTION VALUE

     1)     0.9484789

 VARIABLE        VALUE          REDUCED COST
       T1        0.000100          0.000000
       T2        0.105187          0.000000
       T3        0.000100          0.000000
       W1        0.047177          0.000000
       W2        0.000100          0.000000
       W3        3.644364          0.000000
```

ROW	SLACK OR SURPLUS	DUAL PRICES
2)	0.058110	0.000000
3)	0.000000	-0.065455
4)	0.000000	-1.210909
5)	0.051521	0.000000
6)	0.000000	-4.974545
7)	0.105087	0.000000
8)	0.000000	-1.145455
9)	0.047077	0.000000
10)	0.000000	0.000000
11)	3.644264	0.000000
12)	0.000000	0.949091

Chapter 7 Solutions

Section 7.1

1.

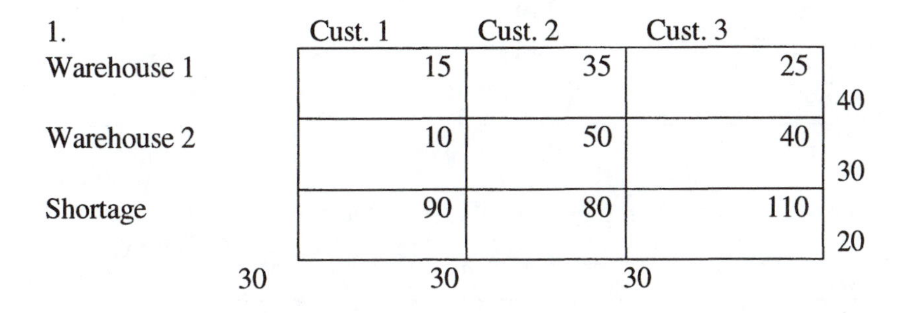

	Cust. 1	Cust. 2	Cust. 3	
Warehouse 1	15	35	25	40
Warehouse 2	10	50	40	30
Shortage	90	80	110	20
	30	30	30	

2.

	C1	C2	C3	DUMMY	
W1	15	35	25	0	40
W2	10	50	40	0	30
W1 EXTRA	115	135	125	0	20
W2 EXTRA	110	150	140	0	20
	30	30	30	20	

Section 7.2

1. By NW corner method we obtain following bfs for Problem 1:

30	10		40
	20	10	30
		20	20
30	30	30	

By NW corner method we obtain the following bfs for Problem 2:

	C1	C2	C3	DUMMY	
W1	15 30	35 10	25	0	40
W2	10	50 20	40 10	0	30
W1 EXTRA	115	135	125 20	0 0	20
W2 EXTRA	110	150	140	0 20	20
	30	30	30	20	

By NW corner method we obtain following bfs for Problem 3:

	M1	M2	M3	M4	M5	M6	DUMMY	
1R	200							200
1O	0	100						100
2R		160	40					200
2O			100					100
3R			100	100				200
3O				100				100
4R				140	60			200
4O					100			100
5R					30	150	20	200
5O							100	100
6R							200	200
6O							100	100
	200	260	240	340	190	150	420	

Section 7.3

1. We begin with the bfs obtained in Section 7.2

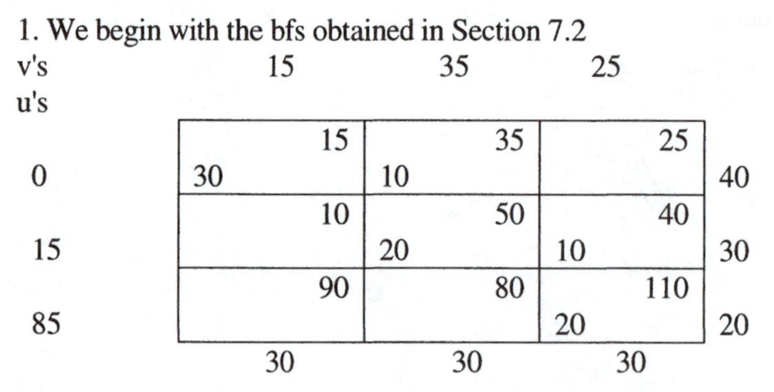

u's \ v's	15	35	25	
0	15 30	35 10	25	40
15	10	50 20	40 10	30
85	90	80	110 20	20
	30	30	30	

Since $\bar{c}_{32}=40$ we enter x_{32} into basis. The loop involving x_{32} and some of the basic variables is (3, 2) - (2, 2) - (2, 3) - (3, 3). x_{33} exits yielding the following bfs

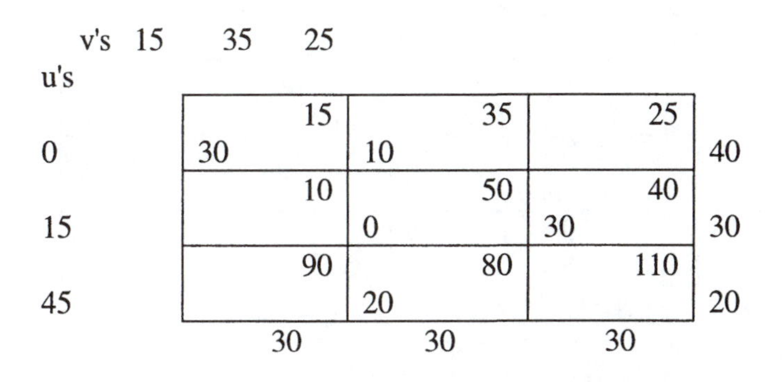

u's \ v's	15	35	25	
	15	35	25	
0	30	10		40
	10	50	40	
15		0	30	30
	90	80	110	
45		20		20
	30	30	30	

Since $\bar{c}_{21}=20$ we enter x_{21} into the basis. The loop involving x_{21} and some of the basic variables is (2,1)-(1,1)-(1,2)-(2,2). After x_{22} exits we obtain the following bfs:

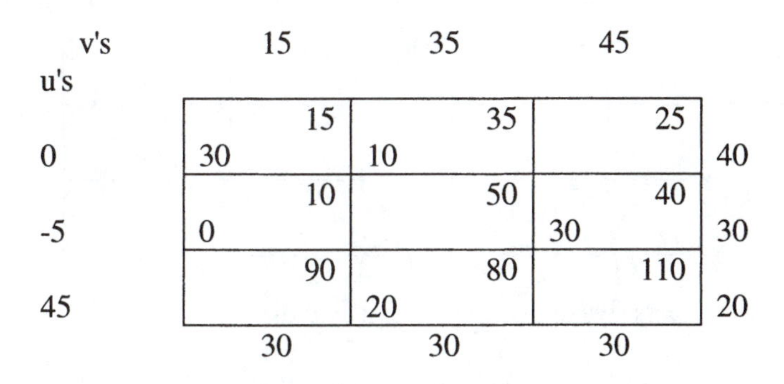

u's \ v's	15	35	45	
	15	35	25	
0	30	10		40
	10	50	40	
-5	0		30	30
	90	80	110	
45		20		20
	30	30	30	

Now $\bar{c}_{13}=20$ so we enter x_{13}. The relevant loop is (1, 3) - (2, 3) - (2, 1) - (1, 1). After x_{11} exits we obtain the following bfs:

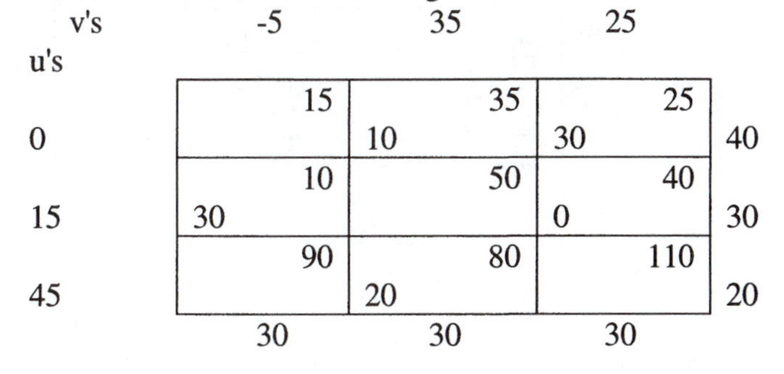

u's \ v's	-5	35	25	
	15	35	25	
0		10	30	40
	10	50	40	
15	30		0	30
	90	80	110	
45		20		20
	30	30	30	

This is an optimal tableau. Thus 10 units should be sent from Warehouse 1 to Customer 2, 30 units from Warehouse 1 to Customer 3, 30 units from Warehouse 2 to customer 1. 20 units of Customer 2's demand will be unsatisfied.

2.

	C1	C2	C3	DUMMY	
W1	15 30	35 10	25	0	40
W2	10	50 20	40 10	0	30
W1 EXTRA	115	135	125 20	0 0	20
W2 EXTRA	110	150	140	0 20	20
	30	30	30	20	

x_{21} enters and x_{22} exits yielding

	C1	C2	C3	DUMMY	
W1	15 10	35 30	25	0	40
W2	10 20	50	40 10	0	30
W1 EXTRA	115	135	125 20	0 0	20
W2 EXTRA	110	150	140	0 20	20
	30	30	30	20	

x_{13} enters and either x_{11} or x_{23} exits (we choose x_{23} to exit) yielding

	C1	C2	C3	DUMMY	
W1	15 0	35 30	25 10	0	40
W2	10 30	50	40	0	30
W1 EXTRA	115	135	125 20	0 0	20
W2 EXTRA	110	150	140	0 20	20
	30	30	30	20	

x_{41} enters yielding the following optimal tableau:

	C1	C2	C3	DUMMY	
W1	15	35 30	25 10	0	40
W2	10 30	50	40	0	30
W1 EXTRA	115	135	125 20	0 0	20
W2 EXTRA	110 0	150	140	0 20	20
	30	30	30	20	

Section 7.4

1. x_{14} is a non-basic variable in the optimal solution. If we change c_{14} to $9 + \Delta$, we find that

$\bar{c}_{14} = 0 + 2 - (9 + \Delta) = -7 - \Delta$.

Thus the current basis remains optimal for $-7 - \Delta \leq 0$ or $\Delta \geq -7$. Thus the current basis remains optimal for $c_{14} \geq 9 - 7 = 2$.

3. x_{23} is a basic variable in the optimal tableau. Thus the new optimal solution leaves all variables the same except for x_{23}, which is increased by 3 to $x_{23} = 5 + 3 = 8$. z increases by 3(13), so the new optimal z-value is 1020 + 39 = 1059.

Section 7.5

1. We use a cost of M to rule out forbidden assignments and add a dummy job (Job 5) to balance the problem. Assigning a person to the dummy job has a 0 cost.

Person \ Job	1	2	3	4	5	Row Min
1	22	18	30	18	0	0
2	18	M	27	22	0	0
3	26	20	28	28	0	0
4	16	22	M	14	0	0
5	21	M	25	28	0	0
Column Min	16	18	25	14	0	

Since all Row minima are 0 we proceed to the column minima. The reduced cost matrix is

Job

Person		1	2	3	4	5
	1	6	0	5	4	0
	2	2	M	2	8	0
Person	3	10	2	3	14	0
	4	0	4	M	0	0
	5	5	M	0	14	0

Only 4 lines (we have used Row 1, Row 4, Col. 3 and Col. 5) are needed to cover all 0's in this matrix. The smallest uncovered element is 2 so we subtract 2 from all uncovered costs and add 2 to all twice covered costs. The resulting matrix is

Job

Person		1	2	3	4	5
	1	6	0	7	4	2
	2	0	M	2	6	0
Person	3	8	0	3	12	0
	4	0	4	M	0	2
	5	3	M	0	12	0

5 lines are needed to cover the 0's so an optimal solution is available: $x_{12}=1$, $x_{21}=1$, $x_{35}=1$, $x_{44}=1$, $x_{53}=1$. Thus Person 3 is not assigned a job and a total time of 18+18+14+25=75 time units is required to complete all the jobs.

4. Cost matrix and optimal assignments (denoted by *) are as follows:

Job

Person	1	2	3	4	Dummy
1	50	46	42	40*	0
2	51*	48	44	1000	0
2’	51	48	44*	1000	0
3	1000	47	45	45	0
3’	1000	47	45	45	0

Note: 1000 rules out prohibited assignment. Total cost = 182.

Section 7.6

2. (Supplies and demands are in thousands) Total Supply-Total Demand=350-300=50 so dummy demand pt. has demand of 50. We obtain the following balanced transportation tableau:

	Mobile	Galv.	NY	LA	Dummy	Supplies
Well 1	10	13	25	28	0	150
Well 2	15	12	26	25	0	200
Mobile	0	6	16	17	0	0+350=350
Galv.	6	0	14	16	0	0+350=350
NY	M	M	0	15	0	0+350=350
LA	M	M	15	0	0	0+350
Demands	350	350	140 +350	160 +350	50	

Chapter 8 Solutions

Section 8.2

1. First label node 1 with a permanent label: [0* 7 12 21 31 44]
   Now node 2 receives a permanent label [0* 7* 12 21 31 44].

Node Temporary Label (* denotes next assigned permanent label)
3 min{12,7+7} = 12*
4 min{21,7+12} = 19
5 min{31,7+21} = 28
6 min{44,7+31} = 38
Now labels are [0* 7* 12* 19 28 38]

Node Temporary Label (* denotes next assigned permanent label)
4 min{19,12+7} = 19*
5 min{28,12+12} = 24
6 min{38,12+21} = 33
Now labels are [0* 7* 12* 19* 24 33]

Node Temporary Label (* denotes next assigned permanent label)
5 min{24,19+7} = 24*
6 min{33, 19+12} = 31
Now labels are [0* 7* 12* 19* 24* 31]
Node Temporary Label (* denotes next assigned permanent label)
6 min{31,24+7} = 31
Now labels are [0* 7* 12* 19* 24* 31*]
31 -24 = c_{56}, 24 - 12 = c_{35}, 12 - 0 = c_{13}. Thus 1-3-6 is the shortest path (of length 31) from node 1 to node 6.

Section 8.3

1. max $z = x_0$
   s.t. $x_{so,1} \le 6$, $x_{so,2} \le 2$ $x_{12} \le 1$, $x_{32} \le 3$, $x_{13} \le 3$, $x_{3,si} \le 2$, $x_{24} \le 7$, $x_{4,si} \le 7$

$x_0 = x_{so,1} + x_{so,2}$ (Node so)
$x_{so,1} = x_{13} + x_{12}$ (Node 1)
$x_{12} + x_{32} + x_{s0,2} = x_{24}$ (Node 2)

$x_{13} = x_{32} + x_{3,si}$ (Node 3)
$x_{24} = x_{4,si}$ (Node 4)
$x_{3,si} + x_{4,si} = x_0$ (Node si)
All variables ≥ 0

Initial flow of 0 in each arc. Begin by labeling sink via path of forward arcs (so, 1) - (1, 3) - (3, 2) - (2, 4) - (4, si). Increase flow in each of these feasible arcs by 3, yielding the following feasible flow:

Arc	Flow
so-1	3
so-2	0
1-3	3
1-2	0
2-4	3
3-si	0
3-2	3
4-si	3

Flow to sink 3

Now label sink by (so-2) - (2-4), (4, si). Each arc is a forward arc and we can increase flow in each arc by 2. This yields the following feasible flow:

Arc	Flow
so-1	3
so-2	2
1-3	3
1-2	0
2-4	5
3-si	0
3-2	3
4-si	5

Flow to sink 5

Now label sink by (so-1) - (1, 2) - (3, 2) - (3, si). All arcs on this path are forward arcs except for (3, 2), which is a backwards arc. We can increase the flow on each forward arc by 1 and decrease the flow on each backward arc by 1. This yields the following feasible flow:

Arc	Flow
so-1	4
so-2	2
1-3	3

1-2	1
2-4	5
3-si	1
3-2	2
4-si	5

Flow to sink 6

The sink cannot be labeled, so we found the maximum flow of 6 units. The minimum cut is obtained from V' = {3, 2, 4, si}. This cut consists of arcs (1, 3), (1, 2), (so, 2) and has capacity of 3 + 1 + 2 = 6 = maximum flow.

4. Maximum flow is 45. Min Cut Set = {1, 3, and si}. Capacity of Cut Set = 20 + 15 + 10 = 45. See Figure.

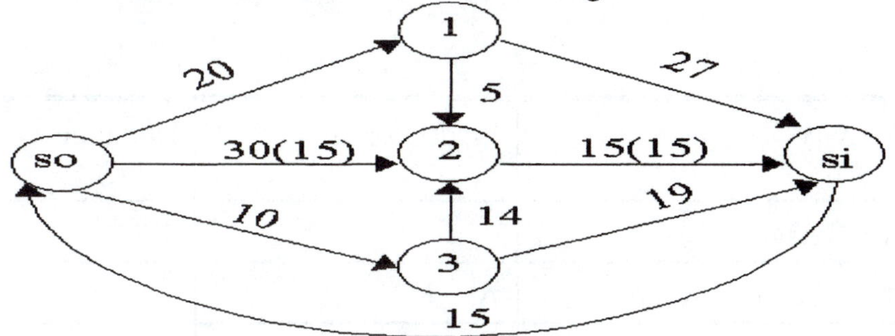

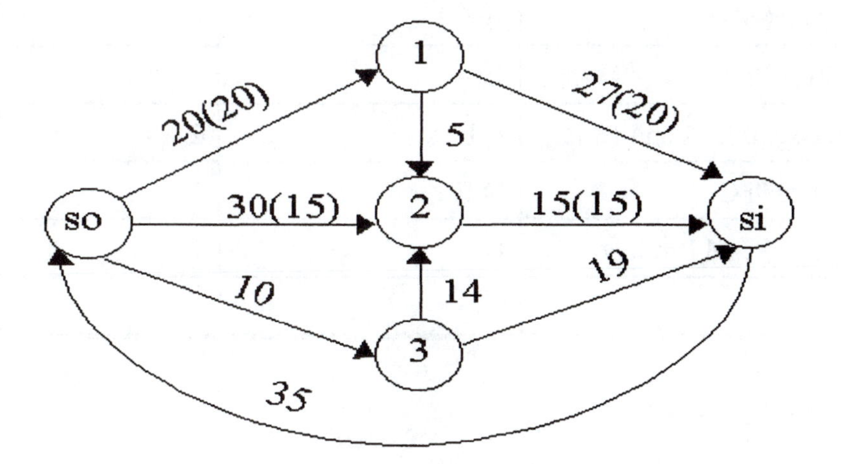

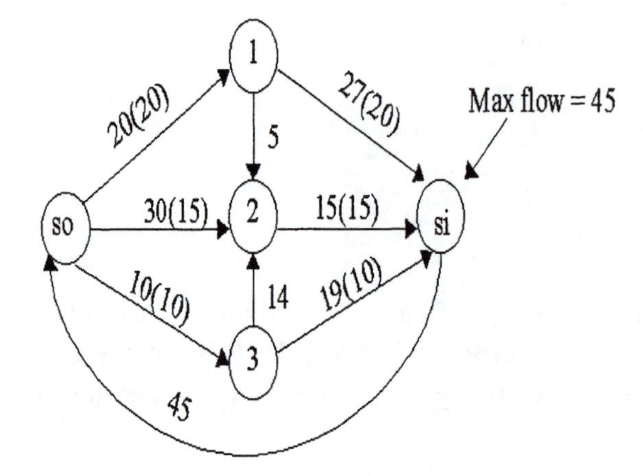

Section 8.4

2.	Activity	Predecessors	Duration (wks.)
	A = Design	-	5
	B = Make Part A	A	4
	C = Make Part B	A	5
	D = Make Part C	A	3
	E = Test Part A	B	2
	F = Assemble A and B	C,E	2
	G = Attach C	D,F	1
	H = Test Final Product	G	1

From the project diagram we find that

ET(1) = 0	LT(1) = 0	TF(1,2) = 0	FF(1,2) = 0
ET(2) = 5	LT(2) = 5	TF(2,3) = 0	FF(2,3) = 0
ET(3) = 9	LT(3) = 9	TF(3,4) = 0	FF(3,4) = 0
ET(4) = 11	LT(4) = 11	TF(2,4) = 1	FF(2,4) = 1
ET(5) = 13	LT(5) = 13	TF(4,5) = 0	FF(4,5) = 0
ET(6) = 14	LT(6) = 14	TF(2,5) = 5	FF(2,5) = 5
ET(7) = 15	LT(7) = 15	TF(5,6) = 0	FF(5,6) = 0
		TF(6,7) = 0	FF(6,7) = 0

Looking at the activities with TF of 0, we find that the critical path is 1-2-3-4-5-6-7 (length 15 days).

The appropriate LP is

$$\min z = x_7 - x_1$$

s.t.
$$x_2 \geq x_1 + 5$$
$$x_3 \geq x_2 + 4$$
$$x_4 \geq x_3 + 2$$
$$x_4 \geq x_2 + 5$$
$$x_5 \geq x_2 + 3$$
$$x_5 \geq x_4 + 2$$
$$x_6 \geq x_5 + 1$$
$$x_7 \geq x_6 + 1$$

all variables urs

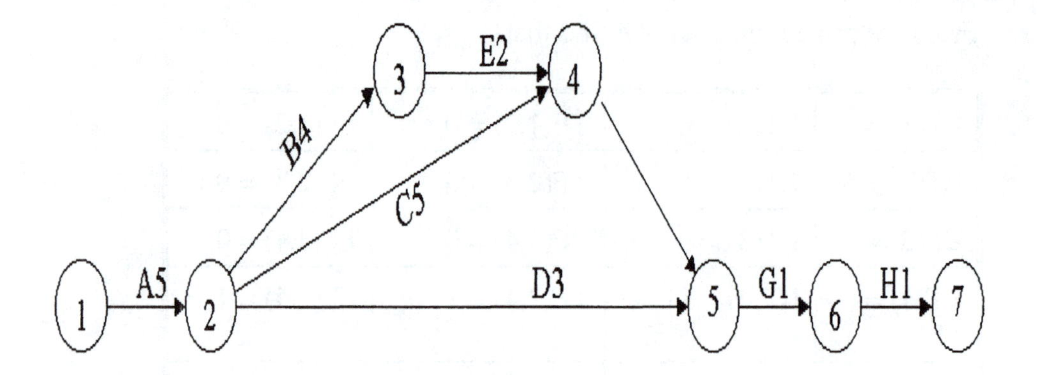

5a.From the project diagram we find that

ET(1) = 0	LT(1) = 0	TF(1,2) = 0	FF(1,2) = 0
ET(2) = 5	LT(2) = 5	TF(2,3) = 0	FF(2,3) = 0
ET(3) = 13	LT(3) = 13	TF(3,5) = 0	FF(3,5) = 0
ET(4) = 17	LT(4) = 17	TF(3,6) = 8	FF(3,6) = 8
ET(5) = 23	LT(5) = 23	TF(3,4) = 0	FF(3,4) = 0
ET(6) = 26	LT(6) = 26	TF(4,5) = 0	FF()4,5 = 0

Both are 1-2-3-5-6 and 1-2-3-4-5-6 are critical paths having length 26 days.

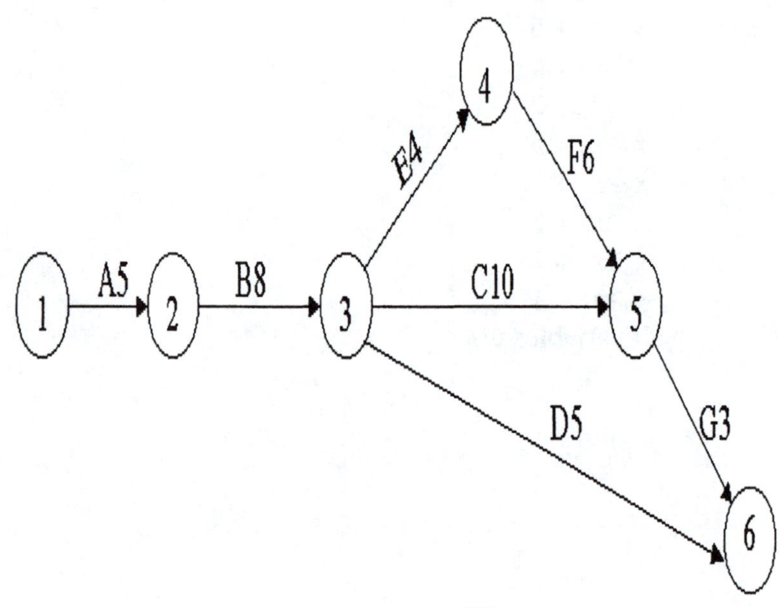

5b. Let A = number of days by which we reduce duration of activity A, etc. and x_j = time that event at node j occurs

min z = 30A + 15B + 20C + 40D + 20E + 30F + 40G

$A \le 2, B \le 3, C \le 1, D \le 2, E \le 2, F \le 3, G \le 1$

$x_2 \ge x_1 + 5 - A$

$x_3 \ge x_2 + 8 - B$

$x_4 \ge x_3 + 4 - E$

$x_5 \ge x_3 + 10 - C$

$x_5 \ge x_4 + 6 - F$

$x_6 \ge x_5 + 3 - G$

$x_6 \ge x_3 + 5 - D$

$x_6 - x_1 \le 20$

$A,B,C,D,E,F,G \ge 0$, x_j urs

Section 8.5

1. min $z = 4x_{12} + 3x_{24} + 2x_{46} + 3x_{13} + 3x_{35} + 2x_{25} + 2x_{56}$

st $x_{12} + x_{13} = 1$ (node 1 constraint)

$x_{12} = x_{24} + x_{25}$ (node 2 constraint)

$x_{13} = x_{35}$ (node 3 constraint)

$x_{24} = x_{46}$ (node 4 constraint)

$x_{25} = x_{56}$ (node 5 constraint)

$x_{46} + x_{56} = 1$ (node 6 constraint)

all $x_{ij} \ge 0$

If $x_{ij} = 1$ the shortest path from node 1 to node 6 will contain arc (i, j) while if $x_{ij} = 0$ the shortest path from node 1 to node 6 does not contain arc (i, j).

3.

Node	Net Outflow
Detroit	6500
Dallas	6000
City 1	-5000
City 2	-4000
City 3	-3000
Dummy	-500

All arcs from Detroit or Dallas to City 1, 2, or 3 have a capacity of 2200. Other arcs have infinite capacity

Arc	Shipping Cost
Detroit-City 1	$2800
Detroit-City 2	$2600
Detroit-City 3	$2300
Detroit-Dummy	$0
Dallas-City 1	$2300
Dallas-City 2	$2000
Dallas-City 3	$2000
Dallas-Dummy	$0

Section 8.6

1. We begin at Gary and include the Gary-South Bend arc. Then we add the South Bend-Fort Wayne arc. Next we add the Gary-Terre Haute arc. Finally we add the Terre Haute-Evansville arc. This minimum spanning tree has a total length of 58 + 79 + 164 + 113 = 414 miles.

Section 8.7

3.) Find the Y values.

$Y_1 = 0, Y_1 - Y_2 = 15, Y_2 - Y_4 = 5, Y_2 - Y_5 = 10, Y_3 - Y_4 = 4$

This yields

$Y_1 = 0, Y_2 = -15, Y_3 = -16, Y_4 = -20, Y_5 = -25$

Find the row 0 coefficients for each non-basic variable.

$\bar{C}_{13} = 0-(-16)-11 = 5$ (satisfies optimality condition)

$\bar{C}_{23} = -15-(-16)-5 = -4$ (satisfies optimality condition)

$\bar{C}_{35} = -16-(-25)-5 = 4$ (violates optimality condition)

$\bar{C}_{45} = -20-(-25)-14 = -9$ (satisfies optimality condition)

Enter X_{35} into the basis.

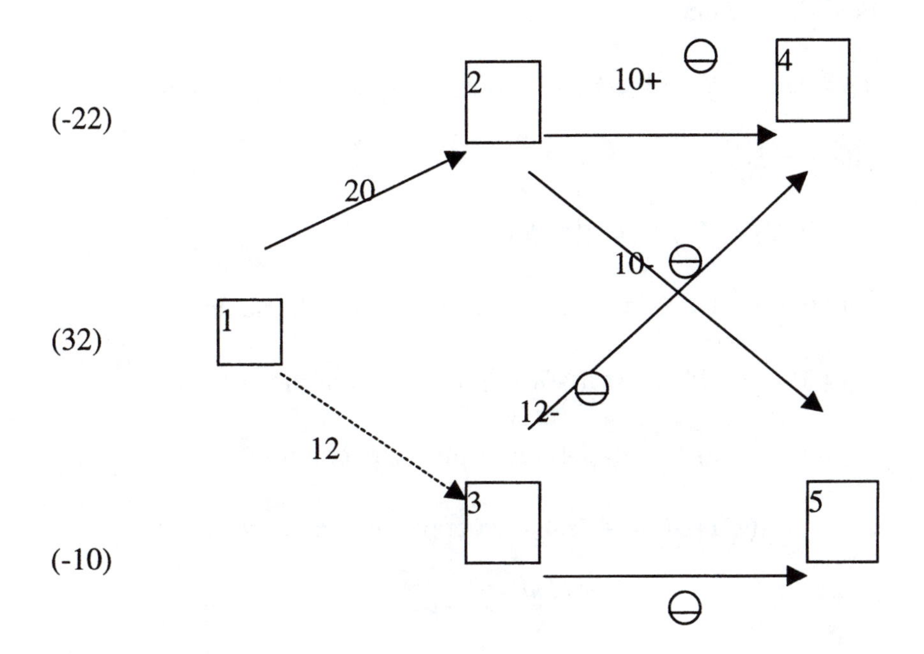

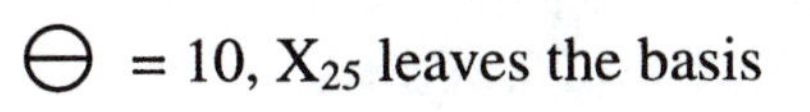

= 10, X_{25} leaves the basis

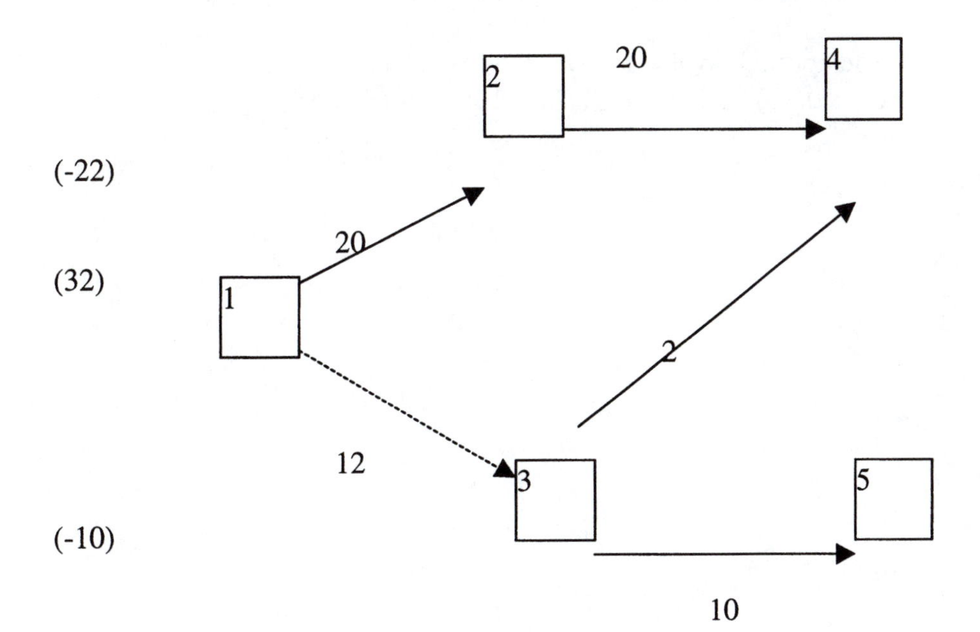

Find the Y values.

$Y_1 = 0$, $Y_1 - Y_2 = 15$, $Y_2 - Y_4 = 5$, $Y_3 - Y_4 = 4$, $Y_3 - Y_5 = 5$

This yields

$Y_1 = 0$, $Y_2 = -15$, $Y_3 = -16$, $Y_4 = -20$, $Y_5 = -21$

Find the row 0 coefficients for each non-basic variable.

$\bar{C}_{13} = 0-(-16)-11 = 5$ (satisfies optimality condition)

$\bar{C}_{23} = -15-(-16)-5 = -4$ (satisfies optimality condition)

$\bar{C}_{25} = -15-(-21)-10 = -4$ (satisfies optimality condition)

$\bar{C}_{45} = -20-(-21)-14 = -13$ (satisfies optimality condition)

Thus, an optimal solution to this MCNFP is
(Basic Variables) $X_{12} = 20$, $X_{24} = 20$, $X_{34} = 2$, $X_{35} = 10$
(Non-basic Variables at upper bound) $X_{13} = 12$
(Non-basic Variables at lower bound) $X_{23} = 0$, $X_{25} = 0$, $X_{45} = 0$

The optimal z-value is
$Z = 20(15) + 12(11) + 20(5) + 2(4) + 10(5) = \$590$

Chapter 9 Solutions

Section 9.2

1. 1. Let $x_i = 1$ if player i starts
   $x_i = 0$ otherwise

Then appropriate IP is

$$\max z = 3x_1 + 2x_2 + 2x_3 + x_4 + 3x_5 + 3x_6 + x_7$$

s.t. $x_1 + x_3 + x_5 + x_7 \geq 4$ (guards)

$x_3 + x_4 + x_5 + x_6 + x_7 \geq 2$ (forwards)

$x_2 + x_4 + x_6 \geq 1$ (center)

$x_1 + x_2 + x_3 + x_4 + x_5 + x_6 + x_7 = 5$

$3x_1 + 2x_2 + 2x_3 + x_4 + 3x_5 + 3x_6 + 3x_7 \geq 10$ (BH)

$3x_1 + x_2 + 3x_3 + 3x_4 + 3x_5 + x_6 + 2x_7 \geq 10$ (SH)

$x_1 + 3x_2 + 2x_3 + 3x_4 + 3x_5 + 2x_6 + 2x_7 \geq 10$ (REB)

$x_6 + x_3 \leq 1$

$-x_4 - x_5 + 2 \leq 2y$

(If $x_1 > 0$ then $x_4 + x_5 \geq 2$)

$x_1 \leq 2(1-y)$

$x_2 + x_3 \geq 1$

$x_1, x_2, \ldots x_7, y$ are all 0-1 variables

3. Let x_1 = Units of Product 1 produced
   x_2 = Units of Product 2 produced
   $y_i = 1$ if any Product i is produced
   $y_i = 0$ otherwise

Then the appropriate IP is

$$\max z = 2x_1 + 5x_2 - 10y_1 - 20y_2$$

s.t. $3x_1 + 6x_2 \leq 120$

$x_1 \leq 40y_1$

$x_2 \leq 20y_2$

$x_1 \geq 0, x_2 \geq 0$, $y_1, y_2 = 0$ or 1

14a. Let $x_i = 1$ if disk i is used, $x_i = 0$ otherwise

$\min z = 3x_1 + 5x_2 + x_3 + 2x_4 + x_5 + 4x_6 + 3x_7 + x_8 + 2x_9 + 2x_{10}$

s.t. $x_1 + x_2 + x_4 + x_5 + x_8 + x_9 \geq 1$ (File 1)

$x_1 + x_3 \geq 1$ (File 2)

$x_2 + x_5 + x_7 + x_{10} \geq 1$ (File 3)

$x_3 + x_6 + x_8 \geq 1$ (File 4)

$x_1 + x_2 + x_4 + x_6 + x_7 + x_9 + x_{10} \geq 1$ (File 5)

$x_i = 0$ or 1 (i=1,2,...10)

14b. If $x_3 + x_5 > 0$, then $x_2 \geq 1$ yields

$1 - x_2 \leq 2y$

$x_3 + x_5 \leq 2(1 - y)$ y=0 or 1

(need M=2 because $x_3 + x_5 = 2$ is possible)

Section 9.3

1.

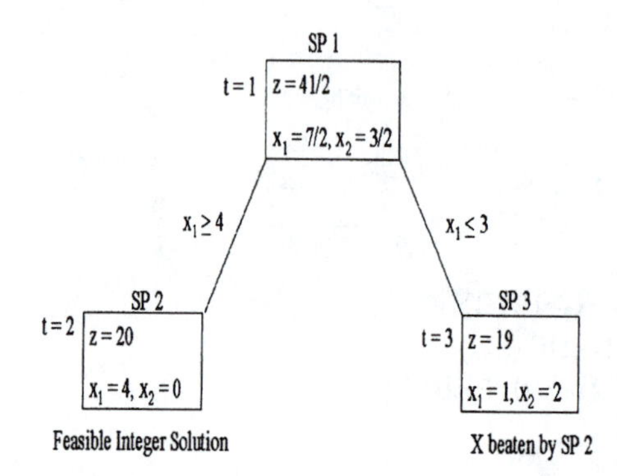

Optimal Solution is $z = 20, x_1 = 4, x_2 = 0$.

2. We wish to solve min $z = 50x_1 + 100x_2$

st $7x_1 + 2x_2 \geq 28$

$2x_1 + 12x_2 \geq 24$

$x_1, x_2 \geq 0$

SP 1

st $7x_1 + 2x_2 \geq 28$

$2x_1 + 12x_2 \geq 24$

$x_1, x_2 \geq 0$

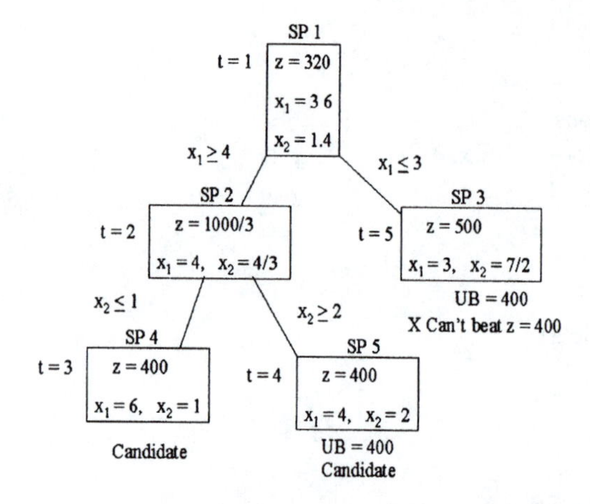

Note: In solving subproblems we have used the result discussed in Problem 8 to conclude that in each subproblem's optimal solution the "newest" constraint must be binding. Thus we know that some optimal solution to SP2 will have $x_1 = 4$.

The two optimal solutions $x_1 = 6$, $x_2 = 1$ and $x_1 = 4$, $x_2 = 2$ have been found.

Section 9.4

3.

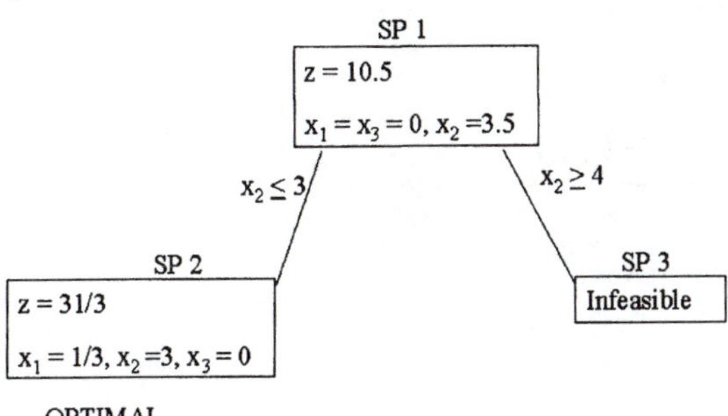

Section 9.5

3.

Letting $x_i = 1$ if item i is chosen and $x_i = 0$ otherwise yields the following knapsack problem:

$$\max z = 5x_1 + 8x_2 + 3x_3 + 7x_4$$
$$\text{st } 3x_1 + 5x_2 + 2x_3 + 4x_4 \leq 6$$
$$x_i = 0 \text{ or } 1$$

We obtain the following tree (for each subproblem any omitted variable equals 0):

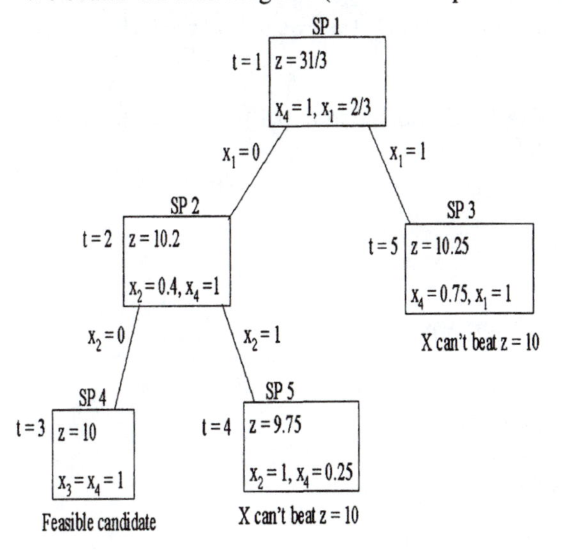

Note that since optimal objective function value for any candidate solution associated with a branch must be an integer, SP 3 can at best yield a z-value of 10, so we need not branch on SP 3. Thus the optimal solution is $z = 10$, $x_1 = x_2 = 0$, $x_3 = x_4 = 1$.

Section 9.6

Let LFR = City 1, LFP = City 2, LR = City 3, and LP = City 4. Then we obtain the following branch and bound tree:

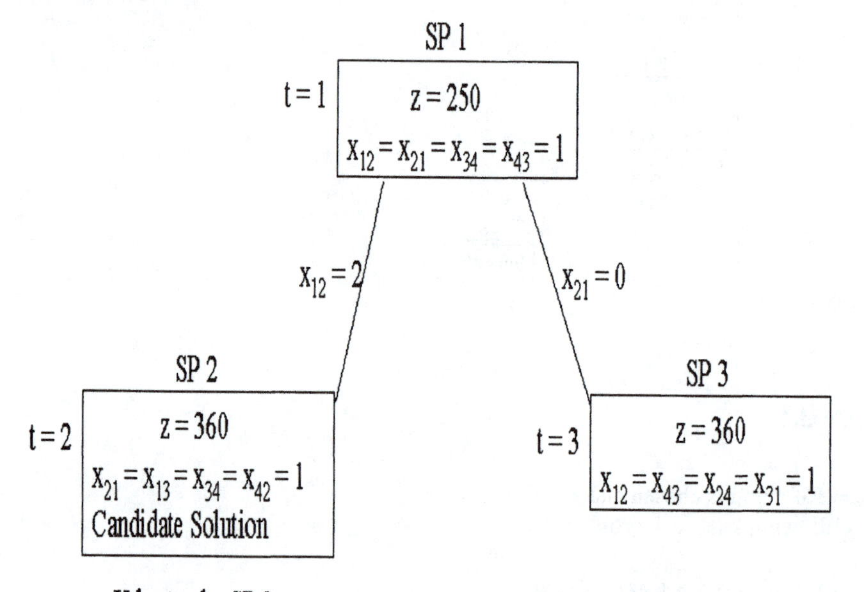

1-2-4-3-1 with total setup time of 330 minutes is optimal. Thus we should produce gasolines in order LFR-LFP-LP-LR-LFR.

3. Add a city 1' and arcs (2,1'), (3,1'), and (4,1'). Cost of arc from (i,1') = cost of arc from (i,1). Now find shortest Hamiltonian path from 1 to 1'.

Section 9.7

3. Let $x_i = 1$ if project i is chosen and $x_i = 0$ otherwise. Then the appropriate IP is:

$$\max z = 5x_1 + 9x_2 + 6x_3 + 3x_4 + 2x_5$$
$$\text{st } 4x_1 + 6x_2 + 5x_3 + 4x_4 + 3x_5 \leq 10$$
$$x_1 + x_2 \leq 1,\ x_3 + x_4 \leq 1,\ x_2 \leq x_5$$
$$\text{All } x_i = 0 \text{ or } 1$$

From the tree we find an optimal solution is $z = 11$, $x_1 = x_3 = 1$, $x_2 = x_4 = x_5 = 0$.

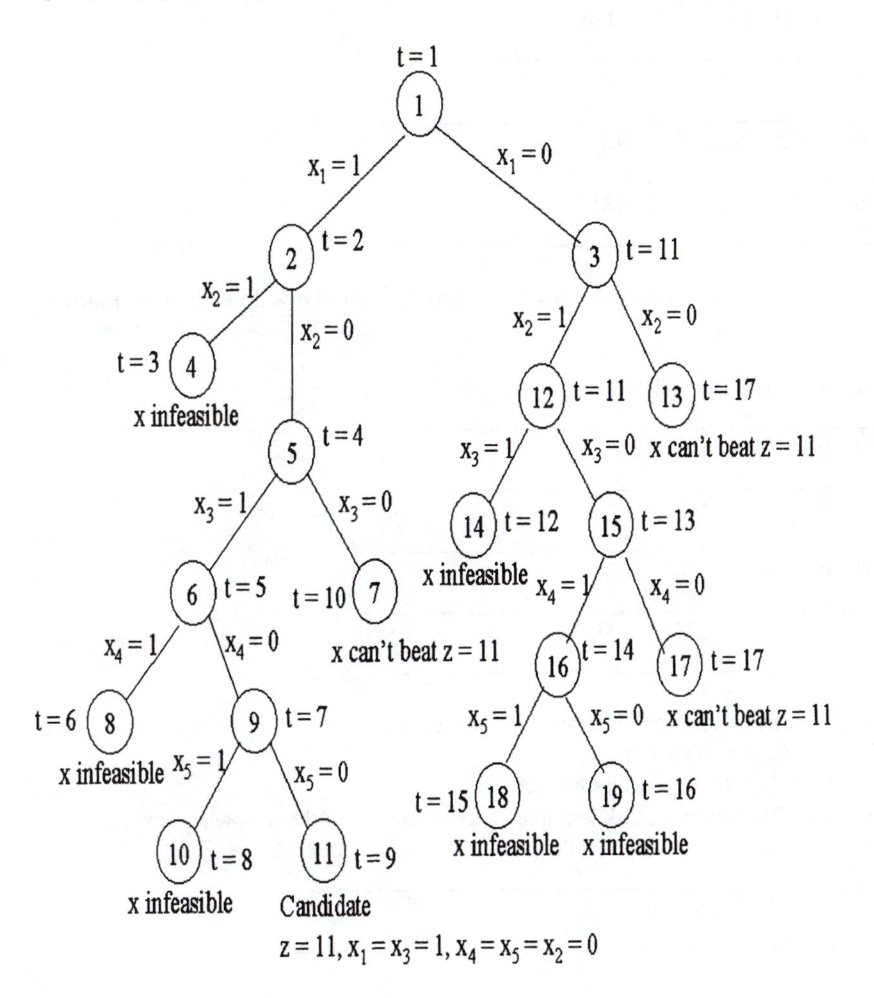

Section 9.8

1. Since both constraints have a fractional part of 1/2 in the optimal tableau, we arbitrarily choose to use the first constraint to yield the cut:

$x_2 + 7s_1/22 + s_2/22 = 3 + 1/2$ or

$x_2 - 3 = 1/2 - 7s_1/22 - s_2/22$ or

$1/2 - 7s_1/22 - s_2/22 \leq 0$. Adding this constraint with a slack variable s_3 yields the following tableau:

z	x_1	x_2	s_1	s_2	s_3	RHS
1	0	0	56/11	30/11	0	126
0	0	1	7/22	1/22	0	7/2
0	1	0	-1/22	3/22	0	9/2
0	0	0	-7/22	-1/22	1	-1/2

The dual simplex pivoting rule indicates that s_1 should enter in row 3 yielding the following tableau:

z	x_1	x_2	s_1	s_2	s_3	RHS
1	0	0	0	2	16	118
0	0	1	0	0	1	3
0	1	0	0	1/7	-1/7	32/7
0	0	0	1	1/7	-22/7	11/7

We now arbitrarily choose row 2 to generate the next cut:
$x_1 + s_2/7 - s_3 + 6s_3/7 = 4/7 + 4$ or
$x_1 - s_3 - 4 = 4/7 - s_2/7 - 6s_3/7$ yielding the cut
$- s_2/7 - 6s_3/7 \leq -4/7$. Adding a slack variable s_4 to the cut yields the following tableau:

z	x_1	x_2	s_1	s_2	s_3	s_4	RHS
1	0	0	0	2	16	0	118
0	0	1	0	0	1	0	3
0	1	0	0	1/7	-1/7	0	32/7
0	0	0	1	1/7	-22/7	0	11/7
0	0	0	0	-1/7	-6/7	1	-4/7

Entering s_2 in the last constraint yields the following (optimal) tableau.

z	x_1	x_2	s_1	s_2	s_3	s_4	RHS
1	0	0	0	0	4	14	110
0	0	1	0	0	1	0	3
0	1	0	0	0	-1	1	4
0	0	0	1	0	-4	1	1
0	0	0	0	1	6	-7	4

This tableau yields the optimal solution $z = 110$ $x_1 = 4$ $x_2 = 3$.

Chapter 10 Solutions

Section 10.1

$\max z = 3x_1 + x_2 + x_3$

s.t. $x_1 + x_2 + x_3 + s_1 = 6$

$2x_1 - x_3 + s_2 = 4$

$x_2 + x_3 + s_3 = 2 \qquad x_1, x_2, x_3 \geq 0$

$BV(0) = \{s_1, s_2, s_3\} \qquad NBV(0) = \{x_1, x_2, x_3\}$

$$B_0^{-1} = B_0 = \begin{bmatrix} 1 & 0 & 0 \\ 0 & 1 & 0 \\ 0 & 0 & 1 \end{bmatrix} \qquad c_{BV} = [0\ \ 0\ \ 0]$$

$$c_{BV} B_0^{-1} = [0\ \ 0\ \ 0]$$

$$\bar{c_1} = [0\ \ 0\ \ 0] \begin{bmatrix} 1 \\ 2 \\ 0 \end{bmatrix} -3 = -3$$

$$\bar{c_2} \bar{=} [0\ \ 0\ \ 0] \begin{bmatrix} 1 \\ 0 \\ 1 \end{bmatrix} \quad -1 = -1$$

$$\bar{c_3} \equiv [0\ \ 0\ \ 0] \begin{bmatrix} 1 \\ -1 \\ 1 \end{bmatrix} -1 = -1$$

$$\text{Column for } x_1 = \begin{bmatrix} 1 & 0 & 0 \\ 0 & 1 & 0 \\ 0 & 0 & 1 \end{bmatrix} \begin{bmatrix} 1 \\ 2 \\ 0 \end{bmatrix} = \begin{bmatrix} 1 \\ 2 \\ 0 \end{bmatrix}$$

$$\text{Right Hand Side} = \begin{bmatrix} 1 & 0 & 0 \\ 0 & 1 & 0 \\ 0 & 0 & 1 \end{bmatrix} \begin{bmatrix} 6 \\ 4 \\ 2 \end{bmatrix} = \begin{bmatrix} 6 \\ 4 \\ 2 \end{bmatrix}$$

$BV(1) = \{s_1, x_1, s_3\}$ $NBV(1) = \{x_2, x_3, s_2\}$

$$B_1^{-1} = \begin{bmatrix} 1 & -1/2 & 0 \\ 0 & 1/2 & 0 \\ 0 & 0 & 1 \end{bmatrix} \qquad c_{BV}B_1^{-1} = [0 \;\; 3 \;\; 0] \begin{bmatrix} 0 & -1/2 & 0 \\ 0 & 1/2 & 0 \\ 0 & 0 & 1 \end{bmatrix} = [0 \;\; 3/2$$

0]

$$\overline{c_2} = [0 \;\; 3/2 \;\; 0] \begin{bmatrix} 1 \\ 0 \\ 1 \end{bmatrix} -1 = -1 \text{ c for } s_2 = 3/2 \quad \text{—}$$

$$\overline{c_3} = [0 \;\; 3/2 \;\; 0] \begin{bmatrix} 1 \\ -1 \\ 1 \end{bmatrix} -1 = -5/2$$

$$\text{Column for } x_3 = \begin{bmatrix} 1 & -1/2 & 0 \\ 0 & 1/2 & 0 \\ 0 & 0 & 1 \end{bmatrix} \begin{bmatrix} 1 \\ -1 \\ 1 \end{bmatrix} = \begin{bmatrix} 3/2 \\ -1/2 \\ 1 \end{bmatrix}$$

$$\text{Right Hand side tableau 1} = \begin{bmatrix} 1 & -1/2 & 0 \\ 0 & 1/2 & 0 \\ 0 & 0 & 1 \end{bmatrix} \begin{bmatrix} 6 \\ 4 \\ 2 \end{bmatrix} = \begin{bmatrix} 4 \\ 2 \\ 2 \end{bmatrix}$$

$BV(2) = \{s_1, x_1, x_3\}$ $\qquad$ $NBV(2) = \{x_2, s_2, s_3\}$

$$B_2^{-1} = \begin{bmatrix} 1 & -1/2 & -3/2 \\ 0 & 1/2 & 1/2 \\ 0 & 0 & 1 \end{bmatrix} \quad c_{BV} B_2^{-1} = [0 \;\; 3 \;\; 1] \begin{bmatrix} 1 & -1/2 & -3/2 \\ 0 & 1/2 & 1/2 \\ 0 & 0 & 1 \end{bmatrix}$$

$=[0 \;\; 3/2 \;\; 5/2]$

$$\overline{c_2} = [0 \;\; 3/2 \;\; 5/2] \begin{bmatrix} 1 \\ 0 \\ 1 \end{bmatrix} - 1 = 3/2, \text{ c for } s_2 > 0, \text{ c for } s_3 > 0$$

so this an optimal tableau.

$$\text{Right Hand Side} = \begin{bmatrix} 1 & -1/2 & -3/2 \\ 0 & 1/2 & 1/2 \\ 0 & 0 & 1 \end{bmatrix} \begin{bmatrix} 6 \\ 4 \\ 2 \end{bmatrix} = \begin{bmatrix} 1 \\ 3 \\ 2 \end{bmatrix}$$

$$\begin{bmatrix} S_1 \\ X_1 \\ X_3 \end{bmatrix} = \begin{bmatrix} 1 \\ 3 \\ 2 \end{bmatrix}$$

$$z = c_{BV} B_2^{-1} b = [0 \;\; 3/2 \;\; 5/2] \begin{bmatrix} 6 \\ 4 \\ 2 \end{bmatrix} = 11$$

Section 10.2

1.

In tableau 0, x_1 enters in row 2 .: $r = 2$, $k = 1$

$$\begin{bmatrix} \overline{a_{11}} \\ \overline{a_{21}} \\ \overline{a_{31}} \end{bmatrix} = \begin{bmatrix} 1 \\ 2 \\ 0 \end{bmatrix}$$

$$E_0 = \begin{bmatrix} 1 & -1/2 & 0 \\ 0 & 1/2 & 0 \\ 0 & 0 & 1 \end{bmatrix} = E_0 B_0^{-1} = B_1^{-1}$$

In tableau 1, x_3 enters in row 3: $r = 3$, $k = 3$

$$\begin{bmatrix} \overline{a_{13}} \\ \overline{a_{23}} \\ \overline{a_{33}} \end{bmatrix} = B_1^{-1} a_3 = \begin{bmatrix} 1 & -1/2 & 0 \\ 0 & 1/2 & 0 \\ 0 & 0 & 1 \end{bmatrix} \begin{bmatrix} 1 \\ -1 \\ 1 \end{bmatrix} = \begin{bmatrix} 3/2 \\ -1/2 \\ 1 \end{bmatrix}$$

$$E_1 = \begin{bmatrix} 1 & 0 & -3/2 \\ 0 & 1 & 1/2 \\ 0 & 0 & 1 \end{bmatrix}$$

$$B_2^{-1} = E_1 B_1^{-1} = \begin{bmatrix} 1 & 0 & -3/2 \\ 0 & 1 & 1/2 \\ 0 & 0 & 1 \end{bmatrix} \begin{bmatrix} 1 & -1/2 & 0 \\ 0 & 1 & 1/2 \\ 0 & 0 & 1 \end{bmatrix} =$$

$$\begin{bmatrix} 1 & -1/2 & -3/2 \\ 0 & 1/2 & 1/2 \\ 0 & 0 & 1 \end{bmatrix}$$

The Rest of problem proceeds as before.

Section 10.3

3.

x_1	3	0	0	3
x_2	2	1	0	1
x_3	2	0	1	0
x_4	0	2	0	3
x_5	0	1	1	2
x_6	0	0	2	1

$$\min z = x_1 + x_2 + x_3 + x_4 + x_5 + x_6$$

s.t. $3x_1 + 2x_2 + 2x_3 \geq 80$

$x_2 + 2x_4 + x_5 \geq 50$

$x_3 + x_5 + 2x_6 \geq 100$

$BV0 = \{x_1, x_4, x_6\}$

$$B_0 = \begin{bmatrix} 3 & 0 & 0 \\ 0 & 2 & 0 \\ 0 & 0 & 2 \end{bmatrix} \qquad B_0^{-1} = \begin{bmatrix} 1/3 & 0 & 0 \\ 0 & 1/2 & 0 \\ 0 & 0 & 1/2 \end{bmatrix}$$

$$c_{BV}B_0^{-1} = [1 \ \ 1 \ \ 1] \begin{bmatrix} 1/3 & 0 & 0 \\ 0 & 1/2 & 0 \\ 0 & 0 & 1/2 \end{bmatrix} = [1/3 \ \ 1/2 \ \ 1/2]$$

$$c_{BV}B_0^{-1}\begin{bmatrix}a_4\\a_6\\a_7\end{bmatrix}-1 = [1/3\ 1/2\ 1]\begin{bmatrix}a_4\\a_6\\a_7\end{bmatrix}-1 = 1/3a_4 + 1/2a_6 + a_7 - 1$$

max $(1/3)a_4 + (1/2)a_6 + (1/2)a_7 - 1$

s.t. $4a_4 + 6a_6 + 7a_7 \le 15$

Optimal solution is z= 1/6 $a_4 = 2$, $a_7 = 1$. Thus x_3 enters the basis.

$$x_3 \text{ column} = B_0^{-1}\begin{bmatrix}2\\0\\1\end{bmatrix} = \begin{bmatrix}1/3 & 0 & 0\\0 & 1/2 & 0\\0 & 0 & 1/2\end{bmatrix}\begin{bmatrix}2\\0\\1\end{bmatrix} = \begin{bmatrix}2/3\\0\\1/2\end{bmatrix}$$

$$\text{Right Hand Side} = B_0^{-1}b = \begin{bmatrix}1/3 & 0 & 0\\0 & 1/2 & 0\\0 & 0 & 1/2\end{bmatrix}\begin{bmatrix}80\\50\\100\end{bmatrix} = \begin{bmatrix}80/3\\25\\50\end{bmatrix}$$

Enter x_3 in Row 1 $\qquad$ $BV(1) = \{x_3, x_4, x_6\}$

$$B_1^{-1} = \begin{bmatrix}1/2 & 0 & 0\\0 & 1/2 & 0\\-1/4 & 0 & 1/2\end{bmatrix}$$

$$c_{BV}B_1^{-1} = [1\ \ 1\ \ 1]\begin{bmatrix}1/2 & 0 & 0\\0 & 1/2 & 0\\-1/4 & 0 & 1/2\end{bmatrix} = [1/4\ 1/2\ \ 1/2]$$

$$[1/4 \quad 1/2 \quad 1]\begin{bmatrix} a_4 \\ a_6 \\ a_7 \end{bmatrix} -1 = (1/4)a_6 + (1/2)a_7 - 1$$

max $z = (1/4)a_6 + (1/2)a_7 - 1$

s.t. $4a_4 + 6a_6 + 7a_7 \leq 15$

Optimal solution to this problem has $z = 0$, so we have solved the LP.

$$\begin{bmatrix} x_3 \\ x_4 \\ x_6 \end{bmatrix} = B_1^{-1} b = \begin{bmatrix} 1/2 & 0 & 0 \\ 0 & 1/2 & 0 \\ -1/4 & 0 & 1/2 \end{bmatrix} \begin{bmatrix} 80 \\ 50 \\ 100 \end{bmatrix} = \begin{bmatrix} 40 \\ 25 \\ 30 \end{bmatrix}$$

$z = 95$

Section 10.4

1.

max $z = 7x_1 + 5x_2 + 3x_3$

s.t. $x_1 + 2x_2 + x_3 \leq 10$ Centralized Constraint

$x_1 \leq 3$ Constraint Set 1

$x_2 + x_3 \leq 5$ Constraint Set 2

$2x_2 + x_3 \leq 8$

$x_1, x_2, x_3 \leq 0$

$[x_1] = u_1 [0] + u_2[3] = [3u_2]$

$$\begin{bmatrix} x_2 \\ x_3 \end{bmatrix} = \lambda_1 \begin{bmatrix} 0 \\ 0 \end{bmatrix} + \lambda_2 \begin{bmatrix} 4 \\ 0 \end{bmatrix} + \lambda_3 \begin{bmatrix} 3 \\ 2 \end{bmatrix} + \lambda_4 \begin{bmatrix} 0 \\ 5 \end{bmatrix} = \begin{bmatrix} 4\lambda_2 + 3\lambda_3 \\ 2\lambda_3 + 5\lambda_4 \end{bmatrix}$$

Objective function:

$7x_1 + 5x_2 + 3x_3 = 7(3u_2) + 5(4\lambda_2 + 3\lambda_3) + 3(2\lambda_3 + 5\lambda_4)$

$= 21u_2 + 20\lambda_2 + 21\lambda_3 + 15\lambda_4$

Centralized Constraint: $3u_2 + 8\lambda_2 + 8\lambda_3 + 5\lambda_4 \leq 10$
Restricted Master:

$$\text{Max } z = 21u_2 + 20\lambda_2 + 21\lambda_3 + 15\lambda_4$$

$$\text{s.t.} \quad 3u_2 + 8\lambda_2 + 8\lambda_3 + 5\lambda_4 + s_1 = 10$$

$$u_1 + u_2 = 1$$

$$\lambda_1 + \lambda_2 + \lambda_3 + \lambda_4 = 1$$

$$u_i, \lambda_i \geq 0$$

$$BV(0) = \{s_1, u_1, \lambda_1\}$$

$$B_0^{-1} = \begin{bmatrix} 1 & 0 & 0 \\ 0 & 1 & 0 \\ 0 & 0 & 1 \end{bmatrix}$$

$$c_{BV} B_0^{-1} = [0 \ 0 \ 0] \begin{bmatrix} 1 & 0 & 0 \\ 0 & 1 & 0 \\ 0 & 0 & 1 \end{bmatrix} = [0 \ 0 \ 0]$$

Obj. Func. Coeff for $u_i = 7x_1$

$$\text{Column in Constraints for } u_i = \begin{matrix} x_1 \\ 1 \\ 0 \end{matrix}$$

$$c_{BV} B_0^{-1} \begin{bmatrix} x_1 \\ 1 \\ 0 \end{bmatrix} -7x_1 = -7x_1$$

Min $z = -7x_1$ s.t. $x_1 \leq 3$, $x_1 \geq 0$
Opt: $z = -21$, $x_1 = 3 \Rightarrow u_2$ enters basis

Obj. Func. Coeff for $\lambda_i = 5x_2 + 3x_3$

$$2x_2 + x_3$$

Col. in Constraints for λ_i $\begin{bmatrix} 0 \\ 1 \end{bmatrix}$

$$c_{BV} B_0^{-1} \begin{bmatrix} 2x_2 + x_3 \\ 0 \\ 1 \end{bmatrix} - (5x_2 + 3x_3) = -5x_2 - 3x_3$$

Min $z = -5x_2 - 3x_3$

s.t. $x_2 + x_3 \leq 5$

$2x_2 + x_3 \leq 8 \qquad x_3, x_2 \geq 0$

Opt: $z = -21$, $x_2 = 3$, $x_3 = 2$ (λ_3)

Arbitrarily choose u_2 to enter. The u_2 column is

$$B_0^{-1} \begin{bmatrix} 3 \\ 1 \\ 0 \end{bmatrix} = \begin{bmatrix} 3 \\ 1 \\ 0 \end{bmatrix} \qquad B_0^{-1} b = \begin{bmatrix} 1 & 0 & 0 \\ 0 & 1 & 0 \\ 0 & 0 & 1 \end{bmatrix} \begin{bmatrix} 10 \\ 1 \\ 1 \end{bmatrix} = \begin{bmatrix} 10 \\ 1 \\ 1 \end{bmatrix}$$

$BV(1) = \{s_1, u_2, \lambda_1\}$

$$B_1^{-1} = \begin{bmatrix} 1 & -3 & 0 \\ 0 & 1 & 0 \\ 0 & 0 & 1 \end{bmatrix}$$

$$c_{BV} B_1^{-1} = [0 \;\; 21 \;\; 0] \begin{bmatrix} 1 & -3 & 0 \\ 0 & 1 & 0 \\ 0 & 0 & 1 \end{bmatrix} = [0 \;\; 21 \;\; 0]$$

Subproblem 1: $c_{BV} B_1^{-1} \begin{bmatrix} x_1 \\ 1 \\ 0 \end{bmatrix} - 7x_1 = 21 - 7x_1$

Min $z = 21 - 7x_1$ s.t. $x_1 \leq 3, x_1 \geq 0$
Opt. $z = 0, x_1 = 3$

Subproblem 2: $c_{BV} B_1^{-1} \begin{bmatrix} 2x_2 + x_3 \\ 0 \\ 1 \end{bmatrix} - 5x_2 - 3x_3 = - 5 x_2 - 3x_3$

Same LP as last phase => enter λ_3

$$B_1^{-1} \begin{bmatrix} 2(3) + 2 \\ 0 \\ 1 \end{bmatrix} = \begin{bmatrix} 1 & -3 & 0 \\ 0 & 1 & 0 \\ 0 & 0 & 1 \end{bmatrix} \begin{bmatrix} 8 \\ 0 \\ 1 \end{bmatrix} = \begin{bmatrix} 8 \\ 0 \\ 1 \end{bmatrix}$$

$$B_1^{-1} b = \begin{bmatrix} 1 & -3 & 0 \\ 0 & 1 & 0 \\ 0 & 0 & 1 \end{bmatrix} \begin{bmatrix} 10 \\ 1 \\ 1 \end{bmatrix} = \begin{bmatrix} 7 \\ 1 \\ 1 \end{bmatrix}$$

$BV(2) = \{\lambda_3, u_2, \lambda_1\}$ $\quad B_2^{-1} = \begin{bmatrix} 1/8 & -3/8 & 0 \\ 0 & 1 & 0 \\ -1/8 & 3/8 & 1 \end{bmatrix}$

$$c_{BV} B_2^{-1} = [21 \;\; 21 \;\; 0] \begin{bmatrix} 1/8 & -3/8 & 0 \\ 0 & 1 & 0 \\ -1/8 & 3/8 & 1 \end{bmatrix} = [21/8 \;\; 105/8 \;\; 0]$$

$$\text{Sub-problem 2:} \;\; c_{BV} B_2^{-1} \begin{bmatrix} 2x_2 + x_3 \\ 0 \\ 1 \end{bmatrix} - 5x_2 - 3x_3 =$$

$21/8\,(2x_2 + x_3) - 5x_2 - 3x_3 = 1/4\, x_2 - 3/8\, x_3$

min $z = 1/4x_2 - 3/8x_3$

s.t. $x_2 + x_3 \leq 5$

$2x_2 + x_3 \leq 8$

$x_2, x_3 \geq 0$

Opt: $z = -15/8$, $x_2 = 0$, $x_3 = 5$

=> Enter λ_4

$$B_2^{-1} \begin{bmatrix} 2(0) + 5 \\ 0 \\ 1 \end{bmatrix} = \begin{bmatrix} 1/8 & -3/8 & 0 \\ 0 & 1 & 0 \\ -1/8 & 3/8 & 1 \end{bmatrix} \begin{bmatrix} 5 \\ 0 \\ 1 \end{bmatrix} = \begin{bmatrix} 5/8 \\ 0 \\ 3/8 \end{bmatrix}$$

$$B_2^{-1} b = \begin{bmatrix} 1/8 & -3/8 & 0 \\ 0 & 1 & 0 \\ -1/8 & 3/8 & 1 \end{bmatrix} \begin{bmatrix} 10 \\ 1 \\ 1 \end{bmatrix} = \begin{bmatrix} 7/8 \\ 1 \\ 1/8 \end{bmatrix}$$

$$BV(3) = \{\lambda_3, u_2, \lambda_4\} \; ; \; B_3^{-1} = \begin{bmatrix} 1/3 & -1 & -5/3 \\ 0 & 1 & 0 \\ -1/3 & 1 & 8/3 \end{bmatrix}$$

$$c_{BV}B_3^{-1} = [21\ \ 21\ \ 15]\begin{bmatrix} 1/3 & -1 & -5/3 \\ 0 & 1 & 0 \\ -1/3 & 1 & 8/3 \end{bmatrix} = [2\ \ 15\ \ 5]$$

Subproblem 2: $c_{BV}B_3^{-1}\begin{bmatrix} 2x_2 + x_3 \\ 0 \\ 1 \end{bmatrix} - 5x_2 - 3x_3 = 5 - x_2 - x_3$

Min $z = 5 - x_2 - x_3$ s.t. $x_2 + x_3 \le 5$
$2x_2 + x_3 \le 8$ $x_2, x_3 \ge 0$

Opt: $z = 0$ $x_2 = 3, x_3 = 2$ or $x_2 = 0, x_3 = 5$

Optimal Solution has been found.

Section 10.5

1.

max $z = 4x_1 + 3x_2 + 5x_3$

s.t. $2x_1 + 2x_2 + x_3 + x_4 \le 9$

$4x_1 - x_2 - x_3 + x_5 \le 6$

$2x_2 + x_3 \le 5$

$x_1 \le 2, x_2 \le 3, x_3 \le 4, x_4 \le 5, x_5 \le 7$

$x_1, x_2, x_3, x_4, x_5 \ge 0$

Initial Tableau:

$z - 4x_1 - 3x_2 - 5x_3 = 0$ $z = 0$

$2x_1 + 2x_2 + x_3 + x_4 + s_1 = 9$ $s_1 = 9$

$4x_1 - x_2 - x_3 + x_5 + s_2 = 6$ $s_2 = 6$

$2x_2 + x_3 + s_3 = 5$ $s_3 = 5$

For x_3: $BN_1 = 4$ $s_1 = 9 - x_3$ ($s_1 \ge 0$ iff $x_3 \le 9$)

$s_2 = 6 + x_3$ ($s_2 \ge 0$ iff $x_3 \ge -6$)

$s_3 = 5 - x_3$ ($s_3 \ge 0$ iff $x_3 \le 5$) => $BN_2 = 5$

replace x_3 with $4 - x_3'$

$z - 4x_1 - 3x_2 + 5x_3' = 20$ $z = 20$

$2x_1 + 2x_2 - x_3' + x_4 + s_1 = 5$ $s_1 = 5$

$4x_1 - x_2 + x_3' + x_5 + s_2 = 10$ $\quad s_2 = 10$
$2x_2 - x_3' + s_3 = 1$ $\quad s_3 = 1$

For x_1: $BN_1 = 2$ $\quad s_1 = 5 - 2x_1$ ($s_1 \geq 0$ iff $x_1 \leq 5/2$)
$s_2 = 10 - 4x_1$ ($s_2 \geq 0$ iff $x_1 \leq 5/2$) => $BN_2 = 5/2$
replace x_1 by $2 - x_1'$

$z + 4x_1' - 3x_2 + 5x_3' = 28$ $\quad z = 28$
$-2x_1' + 2x_2 - x_3' + x_4 + s_1 = 1$ $\quad s_1 = 1$
$-4x_1' - x_2 + x_3' + x_5 + s_2 = 2$ $\quad s_2 = 2$
$2x_2 - x_3' + s_3 = 1$ $\quad s_3 = 1$

For x_2: $BN_1 = 3$ $\quad s_1 = 1 - 2x_2$ ($s_1 \geq 0$ iff $x_2 \leq 1/2$)
$s_2 = 2 + x_2$ ($s_2 \geq 0$ iff $x_2 \geq -2$)
$s_3 = 1 - 2x_2$ ($s_1 \geq 0$ iff $x_2 \leq 1/2$) => $BN_2 = 1/2$
Enter x_2 in Row 3
$z + 4x_1' + 7/2x_3' + 3/2s_3 = 29\ 1/2$ $\quad z = 210.5$
$-2x_1' + x_4 + s_1 - s_3 = 0$ $\quad s_1 = 1$
$-4x_1' + 1/2x_3' + x_5 + s_2 + 1/2s_3 = 5/2$ $\quad s_2 = 5/2$
$x_2 - 1/2x_3' + 1/2s_3 = 1/2$ $\quad x_2 = 1/2$

Optimal Solution:
$z = 29.5$
$s_1 = 0, s_2 = 5/2, x_2 = 1/2, s_3 = 0, x_1' = 0, x_3' = 0, x_4 = 0, x_5 = 0$
$x_1 = 2 - x_1' = 2 - 0 = 2$
$x_3 = 4 - x_3' = 4 - 0 = 4$

Section 10.6

1. Choose $\varepsilon = .1$. $x^0 = [1/3\ 1/3\ 1/3]^T$ and $k = 0$. Since $z = 2/3 > .1$, we proceed to Step 3.

$A = [1\ 0\ -1]$ $\qquad$ $\text{Diag } x^0 = \begin{bmatrix} 1/3 & 0 & 0 \\ 0 & 1/3 & 0 \\ 0 & 0 & 1/3 \end{bmatrix}$

$A[\text{Diag}(x^0)] = [1/3\ 0\ -1/3]$

$$P = \begin{bmatrix} 1/3 & 0 & -1/3 \\ 1 & 1 & 1 \end{bmatrix}$$

$$PP^T = \begin{bmatrix} 2/9 & 0 \\ 0 & 3 \end{bmatrix}$$

$$(PP^T)^{-1} = \begin{bmatrix} 9/2 & 0 \\ 0 & 1/3 \end{bmatrix}$$

$$(I - P^T(PP^T)^{-1}P) = \begin{bmatrix} 1/6 & -1/3 & 1/6 \\ -1/3 & 2/3 & -1/3 \\ 1/6 & -1/3 & 1/6 \end{bmatrix}$$

$c = [1\ 2\ -1]$

$[\text{Diag } x^0]c^T = [1/3\ 2/3\ -1/3]^T$

$(I - P^T (PP^T)^{-1}P)[\text{Diag } x^0]c^T = [-2/9\ 4/9\ -2/9]^T$

Using $\theta = .25$ we obtain $y^1 = [3/8\ 1/4\ 3/8]^T$ and $x^1 = [3/8\ 1/4\ 3/8]^T$. Since $z = .5 > .1$, another iteration would be necessary.

Chapter 11 Solutions

Section 11.1

1.

			Row Min
	2	2	2
	1	3	1
Column Max	2	3	

Since max (row min) = min (column max) = 2, row choosing row 1 and column choosing column 1 is a saddle point. Value of the game is 2 units to the row player.

Section 11.2

Since the third column is dominated by either of the first two columns we need only solve the following game:

1	2
2	0

This game has no saddle point. Let x_1 = probability that row chooses row 1 and y_1 = probability that column chooses column 1.
Row player wants to choose x_1 to maximize

$\min(x_1 + 2(1 - x_1), 2x_1) = \min(2 - x_1, 2x_1)$. This maximum occurs where lines $y = 2 - x_1$ and $y = 2x_1$ intersect (at $x_1 = 2/3$), Value of game to row player is $2 - (2/3) = 4/3$.

Column player chooses y_1 to minimize
$\max(y_1 + 2(1 - y_1), 2y_1) = \max(2 - y_1, 2y_1)$. The minimum occurs where $2 - y_1 = 2y_1$ or $y_1 = 2/3$. Thus value of game to row player is 4/3 and row player's optimal strategy is (2/3, 1/3) and the column player's optimal strategy is (2/3, 1/3, 0).

6. We have a constant-sum game which may be solved as if it is a zero-sum game with the following reward matrix:

		Firm 2 Low	Firm 2 High
Firm 1	Low	500	400
	High	300	600

We find that there is no saddle point so we let L = probability that firm 1's production level is low. Against Firm 2's Low production Firm 1's expected reward = 500L + 300(1 - L) = 200L + 300. Against Firm 2's high production strategy Firm 1's expected reward = 400L + 600(1 - L) = 600 - 200L. By the fundamental assumption of game theory, Firm 1 will receive min[600 - 200L, 200L + 300]. This expression is maximized when 600 - 200L = 200L + 300 or L = 3/4. This yields value of game = $450 to Firm 1.

Let L' = probability that firm 2 chooses low production. Then Firm 2 wishes to minimize Firm 1's expected reward. Against Firm 1's low production level Firm 1 gains 500L' + 400(1 - L') = 400 +100L'. Against Firm 1's high production strategy, Firm 1's expected reward is 300L' + 600(1 - L') = 600 - 300L'. By the fundamental assumption of game theory, Firm 2's choice of L' will result in an expected reward to Firm 1 = max [100L' + 400, 600 - 300L']. This maximum occurs for L' = .5. Thus Firm 1 should choose low production level 3/4 of time while Firm 2 should choose a low production level only 1/2 the time.

Section 11.3

1a.

	Foxhole 1	Foxhole 2	Foxhole 3	Foxhole 4	Foxhole 5
Row 1 A	1	1	0	0	0
Row 2 B	0	1	1	0	0
Row 3 C	0	0	1	1	0
Row 4 D	0	0	0	1	1

1b. Column 2 is dominated by column 1 while column 4 is dominated by column 5.

1c. Suppose gunner chooses strategy which shoots at A, C, and D, 1/3 of the time. Then (assuming that the soldier plays his optimal strategy)no matter where the soldier hides, there is a 1/3 chance that the soldier will be killed. Thus the value to the gunner is 1/3.

1d. If the soldier chooses the given non-optimal strategy, and the gunner always fires at A, the gunner will earn an expected reward of 1/2>1/3.

1e. Row 1 = A,... Row 4 = D

Gunner's LP
max v

st. $v \leq x_1$
$v \leq x_1 + x_2$
$v \leq x_2 + x_3$
$v \leq x_4$, $v \leq x_3 + x_4$
$x_1 + x_2 + x_3 + x_4 = 1$
$x_1, x_2, x_3, x_4 \geq 0$

Soldier's LP
min w

st $w \geq y_1 + y_2$
$w \geq y_2 + y_3$
$w \geq y_3 + y_4$
$w \geq y_4 + y_5$
$y_1 + y_2 + y_3 + y_4 + y_5 = 1$
$y_1, y_2, y_3, y_4, y_5 \geq 0$

For soldier $y_1 = y_3 = y_5 = w = 1/3$ is feasible. For gunner $x_1 = x_3 = x_4 = v = 1/3$ is also feasible. Since these solutions have v = w, they are both optimal strategies.

Section 11.4

3. Rewards are in millions. We assume that each borough will support its own bond issue (this dominates not supporting your own bond issue). Then the reward matrix is as follows:

	Brooklyn Strategies	
Manhattan	Support Manhattan	Oppose Manhattan
Support Brooklyn	(8,8)	(-1,9)
Oppose Brooklyn	(9,-1)	(0,0)

This is a Prisoner's dilemma game with (0, 0) being an equilibrium point. "Oppose" is the non-cooperative action while "Support" is the cooperative action.

Section 11.7

Let (x_1, x_2, x_3, x_4) be a point in the core. Then x_1, x_2, x_3, and x_4 must satisfy

(1) $x_1 + x_2 + x_3 \geq 75$ (2) $x_1 + x_2 + x_4 \geq 75$
(3) $x_1 + x_3 + x_4 \geq 75$ (4) $x_2 + x_3 + x_4 \geq 75$
(5) $x_1 + x_2 + x_3 + x_4 \geq 100$ (6) $x_3 + x_4 \geq 60$
$x_i \geq 0$.

Since (x_1, x_2, x_3, x_4) must be an imputation we also require that (7) $x_1 + x_2 + x_3 + x_4 = 100$. Adding (1)-(4) yields $3(x_1 + x_2 + x_3 + x_4) \geq 300$ or $x_1 + x_2 + x_3 + x_4 \geq 100$. By (7) we now know that (1)- (4) must all hold with equality. Thus any point in the core must satisfy $x_1 = x_2 = x_3 = x_4 = 25$. This, however, violates (6), so the core is empty.

2. Inequality (6) in solution to Problem 1 is now $x_3 + x_4 \geq 50$.
Since (25, 25, 25, 25) satisfies this inequality the core is the point (25, 25, 25, 25).

Chapter 12 Solutions

Section 12.1

1. $\lim_{h \to 0} \frac{3h + h^2}{h} = \lim_{h \to 0} (3 + h) = 3$

3a. $x(-e^{-x}) + e^{-x}$

4. $\partial f/\partial x_1 = 2x_1\exp(x_2)$ $\partial f/\partial x_2 = x_1^2\exp(x_2)$ $\partial f^2/\partial x_1\partial x_2 =$
$\partial f^2/\partial x_2\partial x_1 = 2x_1\exp(x_2) = \partial^2 f/\partial^2 x_1 = 2\exp(x_2), \partial^2 f/\partial^2 x_2 = x_1^2\exp(x_2)$.

Section 12.2

1a. Let S = soap opera ads and F = football ads. Then we wish to

$\min z = 50S + 100F$

st $5S^{1/2} + 17F^{1/2} \geq 40$ (men)

$20S^{1/2} + 7F^{1/2} \geq 60$ (women)

$S \geq 0, F \geq 0$

1b. Since doubling S does not double the contribution of S to each constraint,we are violating the proportionality assumption. Additivity is not violated.

1c. This accounts for the fact that an extra soap opera ad yields a benefit which is a decreasing function of the number of football ads. This accounts for the fact that we may not want to double count people who see both types of ads.

4. Let S = number of soap opera ads and F = number of football ads. See LINGO printout for solution.

LINGO Printout for Problem 4 Section 12.2

```
MODEL:
MIN=50*S+100*F;
5*S^.5+17*F^.5>40;
20*S^.5+7*F^.5>60;
S>0;
F>0;
END
Local optimal solution found at step:                 21
 Objective value:                               563.0744

                       Variable           Value
Reduced Cost
                              S        5.886590
0.0000000
                              F        2.687450
0.0000000

                            Row     Slack or Surplus
Dual Price
                              1        563.0744
1.000000
                              2       0.0000000          -
15.93120
                              3       0.0000000          -
8.148348
                              4        5.886590
0.0000000
                              5        2.687450
0.0000000
```

Section 12.3

2. f"(x)>0 for x>0 and f"(x)<0 for x<0, so f(x) is neither convex nor concave.

5. $f''(x) = -x^{-2} < 0$, so f(x) is a concave function on S.

10. $H = \begin{bmatrix} 2a & b \\ b & 2c \end{bmatrix}$

The function will be convex if $2c \geq 0$, $a \geq 0$, and $4ac \geq b^2$. These conditions ensure that all principal minors will have nonnegative determinants. The function will be concave if $a \leq 0$, $2c \leq 0$ and

$4ac - b^2 \geq 0$. These conditions ensure that both principal minors are nonpositive and the second principal minor is nonnegative.

Section 12.4

1. Let f(x) = profit if $x is spent on advertising. Then
$f(0) = 0$ and for $x>0$, $f(x) = 300x^{1/2} - 100x^{1/2} - 5000 - x$.
Since f(x) has no derivative at x = 0, maximum profit occurs either for x = 0 or a point where $f'(x) = 0$.
Now for $x>0$ $f'(x) = 100x^{-1/2} - 1 = 0$ for x = 10,000. Also
$f''(x) = -50x^{-3/2}<0$ for $x>0$. Thus x = 10,000 is a local maximum(and a maximum over all $x>0$). We now compare f(0) and f(10,000) to determine what the company should do. f(0) = 0 and f(10,000) = $5,000, so company should spend $10,000 on advertising.
If fixed cost is $20,000, $f'(x) = 0$ still holds for x = 10,000. Comparing f(0) = 0 and f(10,000) = -10,000, we now find that x = 0 is optimal.

6. $f'(x) = 3x^2 - 6x + 2$, $f''(x) = 6x - 6$.
The quadratic formula yields $f'(.42) = f'(1.58) = 0$. Since $f''(1.58)>0$ and $f''(.42)<0$ we know that x = 1.58 is a local minimum. Thus extremum candidates are -2, 4, and 1.58. $f'(-2)>0$, so x = -2 is a local minimum. $f'(4)>0$, so x = 4 is a local maximum. Since f(-2) = -25 and f(1.58) = -1.39 we find that the NLP is solved by x = -4.

Section 12.5

1. a = -3 b = 5, so b - a = 8. $x_1 = 5 - .618(8) = .056$
$x_2 = -3 + .618(8) = 1.944$. $f(x_1) = .115$, $f(x_2) = 7.67$
$f(x_2)>f(x_1)$ so interval of uncertainty is now (.056,5]. Then
$x_3 = x_2$, $x_4 = .056 + .618(4.944) = 3.11$.
$f(x_4) = 15.89>f(x_3) = f(x_2) = 7.67$. Thus new interval of uncertainty is (1.944,5]. Now $x_5 = x_4 = 3.11$ and
$x_6 = 1.944 + .618(3.056) = 3.83$. $f(x_5) = 15.89$, $f(x_6) = 22.33$.
Since $f(x_6)>f(x_5)$ new interval of uncertainty is (3.11,5].
$x_7 = x_6 = 3.83$ and $x_8 = 3.11 + .618(1.89) = 4.28$.
$f(x_8) = 26.88$ and $f(x_8)>f(x_7)$. Thus new interval of uncertainty is (3.83,5]. $x_9 = x_8 = 4.28$ and $x_{10} = 3.83 + .618(1.17) = 4.55$.

$f(x_{10}) = 29.8 > f(x_9)$ so new interval of uncertainty is (4.28,5]. This interval has length .72<.8. Thus we know that maximum occurs for some value of x on interval (4.28,5]. (maximum actually occurs for x = 5).

Section 12.6

3.Let $f(q_1, q_2)$ = profit when company 1 sells q_1 units and company 2 sells q_2 units. Then

$$f(q_1, q_2) = (q_1 + q_2)(200 - q_1 - q_2) - q_1 - .5q_2^2$$
$$= 199q_1 + 200q_2 - q_1^2 - 2q_1q_2 - 1.5q_2^2$$

$\partial f/\partial q_1 = 199 - 2q_1 - 2q_2$ and $\partial f/\partial q_2 = 200 - 3q_2 - 2q_1$.
$\partial f/\partial q_1 = \partial f/\partial q_2 = 0$ for $q_1 = 98.5$ and $q_2 = 1$. Since

$$H = \begin{bmatrix} -2 & -2 \\ -2 & -3 \end{bmatrix}$$ det $H_1 = -2<0$, det $H_2 = 2>0$.

Thus the above values of q_1 and q_2 are a local max. Since $f(q_1, q_2)$ is a concave function, we know that (98.5, 1) maximizes profit over all values of q_1 and q_2.

Section 12.7

2. $\nabla f(x_1, x_2) = [3 - 2x_1 \quad -2x_2]$
$\nabla(2.5, 1.5) = [-2, -3]$. To find a new point solve
$\max_{t\geq 0} \; -(0.5 - 2t)^2 - (2.5 - 2t) - (1.5 - 3t)^2 = f(t)$

$f'(t) = 4(0.5 - 2t) + 2 + 6(1.5 - 3t) = 0$ if
$13 - 26t = 0$ or $t = .50$

New Point = (2.5, 1.5) + .5 (-2, -3) = (1.5,0)
Since $\nabla(1.5, 0) = [0 \; 0]$ we conclude the algorithm.

Section 12.8

2. We wish to maximize $L^{2/3}K^{1/3}$ subject to $2L + K = 10$. It is easier to maximize $\ln L^{2/3}K^{1/3} = (2/3)\ln L + (1/3)\ln K$

(this is a concave function so we know that Lagrange multipliers will yield a maximum). Forming the Lagrangian LAG we find that

$LAG = (2/3)\ln L + (1/3)\ln K + \lambda(10 - 2L - K)$
(1) $\partial LAG/\partial L = 2/3L - 2\lambda = 0$, (2) $\partial LAG/\partial K = 1/3K - \lambda = 0$
(3) $\partial LAG/\partial \lambda = 10 - 2L - K = 0$ From (1) $L = 1/3\lambda$.
From (2) $K = 1/3\lambda$ and from (3) $2(1/3\lambda) + (1/3\lambda) = 10$ or
$\lambda = 1/10$. Then $L = K = 10/3$.

3. min $2L + K$
st $L^{2/3}K^{1/3} = 6$ (or $(2/3)\ln L + (1/3)\ln K = \ln 6$)
The Lagrangian (LAG) is given by
$LAG = 2L + K + \lambda(\ln 6 - (2/3)\ln L - (1/3)\ln K)$

(1) $\partial LAG/\partial L = 2 - 2\lambda/3L = 0$, (2) $\partial LAG/\partial K = 1 - \lambda/3K = 0$
(3) $\partial LAG/\partial \lambda = (2/3)\ln L + (1/3)\ln K - \ln 6 = 0$
(1) and (2) yield $L = K = \lambda/3$. Then (3) yields $\lambda = 18$, and (1) and (2) yield $L = K = 6$.

Section 12.9

1. max $p_1(60 - .5p_1) + p_2(40 - p_2) - 10c$
st $60 - .5p_1 - c \le 0$
$40 - p_2 - c \le 0$
All variables ≥ 0
(c = generating capacity)
Ignoring the non-negativity restrictions the K-T conditions consist of the original constraints and
(1) $60 - p_1 + .5\lambda_1 = 0$ (p_1 constraint)
(2) $40 - 2p_2 + \lambda_2 = 0$ (p_2 constraint)
(3) $-10 + \lambda_1 + \lambda_2 = 0$ (c constraint)
(4) $\lambda_1(.5p_1 + c - 60) = 0$
(5) $\lambda_2(p_2 + c - 40) = 0$ $\lambda_1, \lambda_2 \ge 0$
K-T conditions have a solution where $\lambda_1 > 0$ and $\lambda_2 = 0$. Then (3) yields $\lambda_1 = 10$. Now (1) yields $p_1 = 65$ and (2) yields $p_2 = 20$. Finally, (4) yields $c = 27.5$. Since objective function is concave and constraints are linear, we have found the optimal solution.

6. Replace the constraint $-x_1 + x_2 = 1$ by $-x_1 + x_2 \leq 1$ and $x_1 - x_2 \leq -1$. Then (29)-(33)' yield the following K-T conditions.

(1) $2(x_1 - 1) - \lambda_1 + \lambda_2 + \lambda_3 \geq 0$ (x_1 constraint)

(2) $2(x_2 - 2) + \lambda_1 - \lambda_2 + \lambda_3 \geq 0$ (x_2 constraint)

(3) $\lambda_1(1 + x_1 - x_2) = 0$

(4) $\lambda_2(-1 - x_1 + x_2) = 0$

(5) $\lambda_3(2 - x_1 - x_2) = 0$

(6) $x_1(2(x_1 - 1) - \lambda_1 + \lambda_2 + \lambda_3) = 0$

(7) $x_2(2(x_2 - 2) + \lambda_1 - \lambda_2 + \lambda_3) = 0$

Try $\lambda_3 > 0$. Then (5) yields $x_1 + x_2 = 2$. We know, however, that $-x_1 + x_2 = 1$. Solving simultaneously yields $x_1 = 1/2$, $x_2 = 3/2$. If we now try $\lambda_1 = \lambda_2 = 0$ (6) and (7) yield $\lambda_3 = 1$. All the K-T conditions are now satisfied. Since the objective function is convex and the constraints are linear, we have found an optimal solution to the NLP.

Section 12.10

1. Let x_i = amount invested in stock. Then variance of portfolio
$= \text{var}(x_1\mathbf{S}_1 + x_2\mathbf{S}_2 + x_3\mathbf{S}_3)$
$= x_1^2\text{var}\,\mathbf{S}_1 + x_2^2\text{var}\,\mathbf{S}_2 + x_3^2\text{var}\mathbf{S}_3 + 2x_1x_2\text{cov}(\mathbf{S}_1,\mathbf{S}_2)$
$+ 2x_1x_3\text{cov}(\mathbf{S}_1,\mathbf{S}_3) + 2x_2x_3\text{cov}(\mathbf{S}_2,\mathbf{S}_3) = .09x_1^2 + .04x_2^2$
$+ .01x_3^2 + .012x_1x_2 - .008x_1x_3 + .010x_2x_3$

Thus the appropriate NLP is

$$\min z = .09x_1^2 + .04x_2^2 + .01x_3^2 + .012x_1x_2 - .008x_1x_3 + .010x_2x_3$$

$$\text{st} \quad \frac{.15x_1 + .21x_2 + .09x_3}{x_1 + x_2 + x_3} \geq .15$$

or $.06x_2 - .06x_3 \geq 0$ or $x_2 - x_3 \geq 0$

$x_1 + x_2 + x_3 = 100$

$x_1, x_2, x_3 \geq 0$

$$\max p_1(4000 - 10p_1 + p_2) + p_2(2000 - 9p_2 + .8p_1)$$

s.t $2(4000 - 10p_1 + p_2) + 3(2000 - 9p_2 + .8p_1) \leq 5000$ (labor)

$3(4000 - 10p_1 + p_2) + 2000 - 9p_2 + .8p_1 \leq 4500$ (chips)

$p_1, p_2 \geq 0$

LINGO yields an optimal solution of z = -999,535, P1 = \$292.81, P2 = \$158.33, M1 = 0, M2 = \$53.81. Thus adding an additional labor hour will not increase revenue, so Fruit would be willing to pay \$0 for an additional hour of labor. If one more chip were available, then revenue is increased by approximately \$53.81, so Fruit would pay up to (slightly less than) \$53.81 for another chip. Total revenue of \$999,535 will be earned.

Section 12.11

1. We know that in the optimal solution x_1 is between 0 and 3 and x_2 is between 0 and 2. Thus we may choose the following grid points: For x_1: 0, 1, 2, 3 For x_2 0, 2/3, 4/3, 2 Then the approximating problem is

$$\min z = \delta_{12} + 4\delta_{13} + 9\delta_{14} + 4\delta_{22}/9 + 16\delta_{23}/9 + 4\delta_{24}$$

$$\text{st} \quad \delta_{12} + 4\delta_{13} + 9\delta_{14} + 2(4\delta_{22}/9 + 16\delta_{23}/9 + 4\delta_{24}) \le 4$$

$$\delta_{12} + 4\delta_{13} + 9\delta_{14} + 4\delta_{22}/9 + 16\delta_{23}/9 + 4\delta_{24} \le 6$$

$$\delta_{11} + \delta_{12} + \delta_{13} + \delta_{14} = 1,\ \delta_{21} + \delta_{22} + \delta_{23} + \delta_{24} = 1$$

Adjacency assumption plus all variables ≥ 0

Our approximating problem yields the following values for x_1 and x_2:

$x_1 = \delta_{12} + 2\delta_{13} + 3\delta_{14}$, $x_2 = 2\delta_{22}/3 + 4\delta_{23}/3 + 2\delta_{33}$

Section 12.12

1. $x^0 = [1/2\ 1/2]^T$ $\nabla f(x,y) = [4 - 4x - 2y, 6 - 2x - 4y]$

$\nabla f(.5,.5) = [1\ 3]$ Find d^0 by solving

$$\max z = d1 + 3d2$$
$$\text{st} \quad d1 + 2d2 \le 2$$
$$d1, d2 \ge 0$$

Optimal solution is $d^0 = [0\ 1]T$ Choose $x^1 = [.5\ .5]^T + t_0[-.5\ .5]^T$

$= [.5 - .5t_0\ .5 + .5t_0]$ where t_0 solves

$\max \quad f(.5 - .5t, .5 + .5t) = 3.5 + t - t^2/2 = g(t).$

$0 \le t \le 1$

Then $g'(t) = 1 - t = 0$ for $t = 1$. Since $g''(t) < 0$, $t_0 = 1$ and

$x^1 = [0\ 1]^T$. Here $z = f(0, 1) = 4$. $\nabla f(x^1) = [2\ 2]$. We find d^1 by solving

$$\max z = 2d1 + 2d2$$
$$\text{st} \quad d1 + 2d2 \le 2$$
$$d1, d2 \ge 0$$

Optimal solution is $d^1 = [2\ 0]^T$. Now $x^2 = [0\ 1]^T + t_1[2\ -1]^T$
$= [2t_1\ 1 - t_1]$ where t_1 solves
$\max_{0 \le t \le 1} f(2t, 1 - t) = 4 + 2t - 6t^2 = h(t)$. Then $h'(t) = 2 - 12t = 0$
for $t = 1/6$. Since $h''(t)<0$, $t_1 = 1/6$ and $x^2 = [1/3\ 5/6]$. At this point $z = 4.17$.

Section 12.13

We first solve LP to maximize profit

$$\begin{aligned} \max z &= 500x1 + 1100x2 - 10(OT) \\ \text{st} \quad 6x1 + 12x2 &\le 200 \\ 8x1 + 20x2 &\le 300 \\ 11x1 + 24x2 &\le 300 + OT \\ x1, x2, OT &\ge 0 \end{aligned}$$

Here x_i = Units of product I produced and OT = hours of overtime used. Optimal solution has Profit = \$16,666.67 and OT = 83.33. We now add constraint OT≤k, and set k = 80, 75, 70, 65, 60, 50, 40, ... 0 and obtain the given tradeoff curve. For instance, when k = 10 optimal z-value is \$14,108.33, so (10, 14,108.33) is on tradeoff curve.

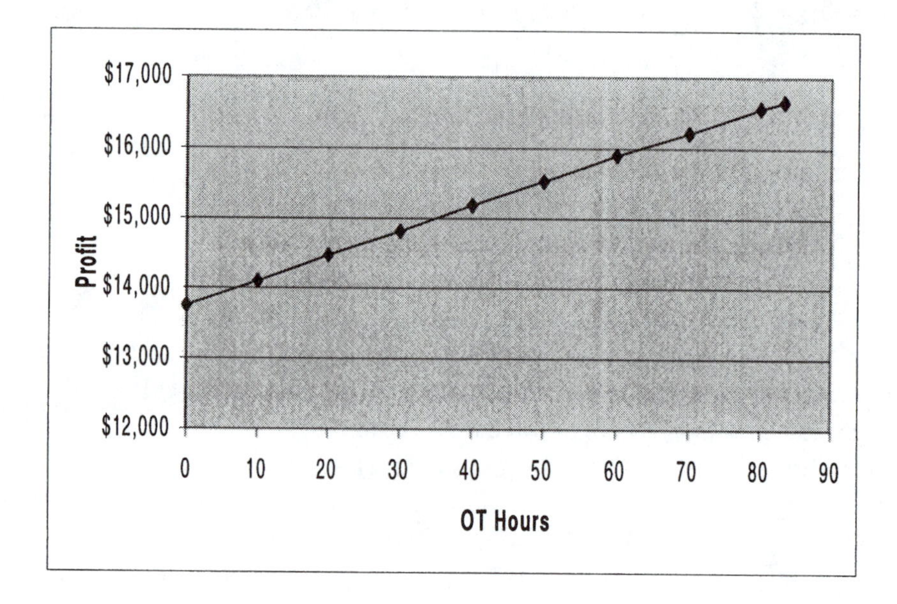

Chapter 13 Solutions

Section 13.1

1. If I can force it to be my opponent's turn with 1 match left, I will win. Working backwards, if I can force my opponent's turn to occur with 6, 11, 16, 21, 26, 31 or 36 matches on the table I will win. Thus I should pick up 40- 36 = 4 matches on the first turn and on each successive turn pick up (5 - # of matches my opponent has picked up on his last turn).

Section 13.2

1. Define $f_t(i)$ to be the shortest path from node i to node 10 given that node is a stage t node.

Define $x_t(i)$ to be the endpoint of the arc that should be chosen if we are in node i.

$f_4(8) = 3 \quad x_4(8) = 10$

$f_4(9) = 4 \quad x_4(9) = 10$

$$f_3(5) = \min\begin{cases} 1 + f_4(8) = 4^* & x_3(5) = 8 \\ 3 + f_4(9) = 7 \end{cases}$$

$$f_3(6) = \min\begin{cases} 6 + f_4(8) = 9 \\ 3 + f_4(9) = 7^* & x_3(6) = 9 \end{cases}$$

$$f_3(7) = \min\begin{cases} 3 + f_4(8) = 6^* & x_3(6) = 8 \\ 3 + f_4(9) = 7 \end{cases}$$

$$f_2(2) = \min\begin{cases} 7 + f_3(5) = 11^* \\ 4 + f_3(6) = 11^* & x_2(2) = 5 \text{ or } 6 \\ 6 + f_3(7) = 12 \end{cases}$$

$$f_2(3) = \min \begin{cases} 3 + f_3(5) = 7^* & x_2(3) = 5 \\ 2 + f_3(6) = 9 \\ 4 + f_3(7) = 10 \end{cases}$$

$$f_2(4) = \min \begin{cases} 4 + f_3(5) = 8^* \\ 1 + f_3(6) = 8^* & x_2(4) = 5 \text{ or } 6 \\ 5 + f_3(7) = 11 \end{cases}$$

$$f_1(1) = \min \begin{cases} 2 + f_2(2) = 13 \\ 4 + f_2(3) = 11^* & x_2(1) = 3 \text{ or } 4 \\ 3 + f_2(4) = 11^* \end{cases}$$

From this analysis we see that 1-3-5-8-10, 1-4-6-9-10,or 1-4-5-8-10 are all shortest paths from node 1 to node 10(each of these paths has length 11).

Section 13.3

2. Let $f_t(i)$ be the minimum cost incurred during months t,t + 1,..3 if the inventory at the beginning of month t is i.
Note that it is clearly suboptimal to produce more than the total demand of 800 units. This means that our inventory at the beginning of month 3 can't exceed 800 - (200 + 300) = 300 units.
Thus we need only compute $f_3(i)$ for i = 0,100,200 and 300. As in the text, during month 3 we simply produce enough to meet month 3 demand from current inventory and production. Let $x_t(i)$ be the quantity that should be produced during month t in order to attain $f_t(i)$. Then

$f_3(300) = 0$ $x_3(300) = 0$
$f_3(200) = 250 + 12(100) = 1450$ $x_3(200) = 100$
$f_3(100) = 250 + 12(200) = 2,650$ $x_3(100) = 200$
$f_3(0) = 250 + 12(300) = 3,850$ $x_3(0) = 300$
Then $f_2(i) = \min\{c(x) + 1.5(i + x\text{-}300) + f_2(i + x\text{-}300)\}$

x
where $c(0) = 0$

$c(100) = 250 + 10(100) = 1{,}250$
$c(200) = 250 + 10(200) = 2{,}250$
$c(300) = 250 + 10(300) = 3{,}250$
$c(400) = 250 + 10(400) = 4{,}250$
$c(500) = 250 + 10(500) = 5{,}250$
$c(600) = 250 + 10(600) = 6{,}250$
$c(700) = 250 + 10(700) = 7{,}250$
$c(800) = 250 + 10(800) = 8{,}250$

and $x \geq 300-i$. Note that during month 2 entering inventory cannot exceed 800 - 200 = 600. Also note that during month 2 it would be foolish to produce more than (300 + 300) - i = 600 - i units, because we would then have some inventory at the end of month 3.
Using these simplifications the necessary $f_2(\ )$ computations are as follows:

$$f_2(0) = \min \begin{cases} 3{,}250 + 0 + f_3(0) = 7100 \text{ (Produce 300)} \\ 4{,}250 + 150 + f_3(100) = 7{,}050 \text{ (Produce 400)} \\ 5{,}250 + 300 + f_3(200) = 7{,}000 \text{ (Produce 500)} \\ 6{,}250 + 450 + f_3(300) = 6{,}700^* \text{ (Produce 600)} \end{cases}$$

$x_2(0) = 600$

$$f_2(100) = \min \begin{cases} 2{,}250 + 0 + f_3(0) = 6{,}100 \text{ (Produce 200)} \\ 3{,}250 + 150 + f_3(100) = 6{,}050 \text{ (Produce 300)} \\ 4{,}250 + 300 + f_3(200) = 6{,}000 \text{ (Produce 400)} \\ 5{,}250 + 450 + f_3(300) = 5{,}700^* \text{(Produce 500)} \end{cases}$$

$x_2(100) = 500$

$$f_2(200) = \min \begin{cases} 1{,}250 + 0 + f_3(0) = 5{,}100 \text{ (Produce 100)} \\ 2{,}250 + 150 + f_3(100) = 5{,}050 \text{ (Produce 200)} \\ 3{,}250 + 300 + f_3(200) = 5{,}000 \text{ (Produce 300)} \\ 4{,}250 + 450 + f_3(300) = 4{,}700^* \text{ (Produce 400)} \end{cases}$$

$x_2(200) = 400$

$$f_2(300) = \min \begin{cases} 0 + 0 + f_3(0) = 3{,}850 \text{ (Produce 0)} \\ 1{,}250 + 150 + f_3(100) = 4{,}050 \text{ (Produce 100)} \\ 2{,}250 + 300 + f_3(200) = 4{,}000 \text{ (Produce 200)} \\ 3{,}250 + 450 + f_3(300) = 3{,}700^* \text{ (Produce 300)} \end{cases}$$

$x_2(300) = 300$

$$f_2(400) = \min \begin{cases} 0 + 150 + f_3(100) = 2{,}800 \text{ (Produce 0)} \\ 1{,}250 + 300 + f_3(200) = 3{,}000 \text{ (Produce 100)} \\ 2{,}250 + 450 + f_3(300) = 2{,}700^* \text{ (Produce 200)} \end{cases}$$

$x_2(400) = 200$

$$f_2(500) = \min \begin{cases} 0 + 300 + f_3(200) = 1{,}750 \text{ (Produce 0)} \\ 1{,}250 + 450 + f_3(300) = 1{,}700^* \text{ (Produce 100)} \end{cases}$$

$x_2(500) = 100$

$f_2(600) = 0 + 450 + f_3(300) = 450 \quad x_2(600) = 0$

Now we compute $f_1(0)$ from

$$f_1(0) = \min_x \{c(x) + 1.5(i + x - 200) + f_2(i + x - 200)\}$$

where $x \geq 200$. Thus

$$f_1(0) = \min \begin{cases} 2{,}250 + 0 + f_2(0) = 8{,}950^* \text{ (Produce 200)} \\ 3{,}250 + 150 + f_2(100) = 9{,}100 \text{ (Produce 300)} \\ 4{,}250 + 300 + f_2(200) = 9{,}250 \text{ (Produce 400)} \\ 5{,}250 + 450 + f_2(300) = 9{,}400 \text{ (Produce 500)} \\ 6{,}250 + 600 + f_2(400) = 9{,}550 \text{ (Produce 600)} \\ 7{,}250 + 750 + f_2(500) = 9{,}700 \text{ (Produce 700)} \\ 8{,}250 + 900 + f_2(600) = 9{,}600 \text{ (Produce 800)} \end{cases}$$

$x_1(0) = 200$

Thus $f_1(0)$ = \$8,950 and $x_1(0)$ = 200 radios should be produced during month 1. This yields a month 2 inventory of 200 - 200 = 0. Thus during month 2 we produce $x_2(0)$ = 600 radios. At the beginning of month 3 the inventory will now be 0 + 600 - 300 = 300. Hence during month 3 $x_3(300)$ = 0 radios should be produced.

Note that the total production cost of this plan is 250 + 200(10) + 250 + 600(10) = 8,500 and Total Holding Cost = 1.5(300) = 450 (assessed on inventory at end of month 2). Thus total cost is 8500 + 450 = \$8,950 = $f_1(0)$.

Section 13.4

2. We use (8). Let g(w) be the maximum benefit that can be attained from a w pound knapsack. Also define x(w) to be the type of item that attains the minimum in the recursion for a w pound knapsack.
Then (Type 0 item means nothing can fit in knapsack)

$g(0) = 0 \quad x(0) = 0$

$g(1) = 0 \quad x(1) = 0$

$g(2) = 2 \quad x(2) = 3$

$$g(3) = \max \begin{cases} 4 + g(0) = 4^* & \text{(Put in Type 2 Item) } x(3) = 2 \\ 2 + g(1) = 2 & \text{(Put in Type 3 Item)} \end{cases}$$

$$g(4) = \max \begin{cases} 5 + g(0) = 5^* & \text{(Put in Type 1 Item) } x(4) = 1 \\ 4 + g(1) = 4 & \text{(Put in Type 2 Item)} \\ 2 + g(2) = 4 & \text{(Put in Type 3 Item)} \end{cases}$$

$$g(5) = \max \begin{cases} 5 + g(1) = 5 & \text{(Put in Type 1 Item)} \\ 4 + g(2) = 6^* & \text{(Put in Type 2 Item) } x(5) = 2 \\ 2 + g(3) = 6^* & \text{(Put in Type 3 Item)} \end{cases}$$

$$g(6) = \max \begin{cases} 5 + g(2) = 7 & \text{(Put in Type 1 Item)} \\ 4 + g(3) = 8^* & \text{(Put in Type 2 Item) } x(6) = 2 \\ 2 + g(4) = 7 & \text{(Put in Type 3 Item)} \end{cases}$$

$$g(7) = \max \begin{cases} 5 + g(3) = 9^* & \text{(Put in Type 1 Item)} \\ 4 + g(4) = 9 & \text{(Put in Type 2 Item) } x(7) = 1 \text{ or } 3 \\ 2 + g(5) = 8^* & \text{(Put in Type 3 Item)} \end{cases}$$

$$g(8) = \max \begin{cases} 5 + g(4) = 10^*\text{(Put in Type 1 Item)} \\ 4 + g(5) = 10^*\text{(Put in Type 2 Item)} \quad x(8) = 2 \\ 2 + g(6) = 10^*\text{(Put in Type 3 Item)} \end{cases}$$

We choose to begin by putting a Type 2 item in the knapsack. This leaves us with an 8 - 3 = 5 pound knapsack so we put an x(5) = Type 2 item in the knapsack. This leaves us with a 5 - 3 = 2 pound knapsack so we next put in an x(2) = Type 3 item to fill the knapsack. Thus the knapsack should be filled with two Type 2 items and one Type 3 item. There are, of course, other optimal solutions such as using two Type 1 items. Observe that we really did not need g(7) to determine g(8)!

Section 13.5

2. Let g(t) be the minimum net cost incurred from Time t to Time 6 given that we have a new car at Time t(including purchase and salvage costs for the car purchased at Time t).

Then g(6) = 0. Also note that

Net Cost of Keeping Car for 1 Year = 10,000 - 7000 + 300 = $3300.

Net Cost of Keeping Car for 2 Years = 10,000 - 6000 + 300 + 500 = $4800.

Net Cost of Keeping Car for 3Years = 10,000 - 4000 + 300 + 500 + 800 = $7600.

Net Cost of Keeping Car for 4 Years = 10,000 -3000 + 300 + 500 + 800 + 1200 = $9800.

Net Cost of Keeping a Car for 5 Years = 10,000 - 2000 + 300 + 500 + 800 + 1200 + 1600 = $12,400.

Net Cost of Keeping a Car for Six Years = 10,000 - 1000 + 300 + 500 + 800 + 1200 + 1600 + 2200 = $15,600.

Then

g(5) = 3300 + g(6) = 3300* (Keep till end of year 6 and trade-in)

$$g(4) = \min \begin{cases} 3300 + g(5) = 6600 \quad \text{(Next trade-in at Time 5)} \\ 4800 + g(6) = 4800^* \quad \text{(Next trade-in at Time 6)} \end{cases}$$

$$g(3) = \min \begin{cases} 3300 + g(4) = 8{,}100 \text{(Next trade-in at Time 4)} \\ 4800 + g(5) = 8{,}100 \text{(Next trade-in at Time 5)} \\ 7600 + g(6) = 7{,}600^* \text{ (Next trade-in at Time 6)} \end{cases}$$

$$g(2) = \min \begin{cases} 3300 + g(3) = 10{,}900 \text{ (Next trade-in at Time 3)} \\ 4800 + g(4) = 9{,}600^* \text{ (Next trade-in at Time 4)} \\ 7600 + g(5) = 10{,}900 \text{ (Next trade-in at Time 5)} \\ 9800 + g(6) = 9{,}800 \text{ (Next trade-in at Time 6)} \end{cases}$$

$$g(1) = \min \begin{cases} 3300 + g(2) = 12{,}900 \text{ (Next trade-in at Time 2)} \\ 4800 + g(3) = 12{,}400^* \text{(Next trade-in at Time 3)} \\ 7600 + g(4) = 12{,}400^* \text{(Next trade-in at Time 4)} \\ 9800 + g(5) = 13{,}100 \text{ (Next trade-in at Time 5)} \\ 12{,}400 + g(6) = 12{,}400^* \text{(Next trade-in at Time 6)} \end{cases}$$

$$g(0) = \min \begin{cases} 3300 + g(1) = 15{,}700 \text{ (Next trade-in at Time 1)} \\ 4800 + g(2) = 14{,}400^* \text{ (Next trade-in at Time 2)} \\ 7600 + g(3) = 15{,}200 \text{ (Next trade-in at Time 3)} \\ 9800 + g(4) = 14{,}600 \text{ (Next trade-in at Time 4)} \\ 12{,}400 + g(5) = 15{,}700 \text{ (Next trade-in at Time 5)} \\ 15{,}600 + g(6) = 15{,}600 \text{ (Next trade-in at Time 6)} \end{cases}$$

Thus the car should be traded in at times 2,4 and 6. A net cost of \$14,400 will be incurred.

Section 13.6

2a. Let d be the amount of money consumed during a year. If $u(d) = d^2$,then $u''(d) = 2>0$. This means that as consumption is increased, each additional dollar of consumption adds more to the person's utility. Most people do not behave this way. See Chapter 13 of OR. On the other hand if $u(d) = \ln d$, then $u''(d) = -1/d^2$, so each additional dollar of consumption adds less and less to Juli's utility. This is more consistent with the behavior of most people. In short, the behavior of few people can be described by convex utility functions.

2b. Let $f_t(d)$ be the maximum utility that can be earned during years t, t + 1,10 given that d dollars are available at the beginning of year t(including

year t income). During year 10 it makes sense to consume all available money (after all there is no future. Thus
$f_{10}(d) = \ln d$

For $t \le 9$
$$f_t(d) = \max_{c} \{\ln c + f_{t+1}(1.1[d - c] + i)\}$$

where $0 \le c \le d$.
We work backwards from the $f_{10}(\ )$'s to $f_1(D)$.

5. Let Subject 1 = French, Subject 2 = English, and Subject 3 = Statistics. Since Probability of passing at least one course equals 1 - (probability of failing all three courses) our problem is equivalent to minimizing the probability of flunking all 3 subjects. Let $f_t(i)$ be the minimum probability of flunking subjects t, t + 1,...3 when i hours can be allocated to these subjects. Let $x_t(i)$ be an optimal allocation of study hours to subject t which attains $f_t(i)$.
Then

$f_3(0) = .90\ x_3(0) = 0$
$f_3(1) = .70\ x_3(1) = 1$
$f_3(2) = .60\ x_3(2) = 2$
$f_3(3) = .55\ x_3(3) = 3$
$f_3(4) = .50\ x_3(4) = 4$
$f_2(0) = .75(.90) = .675^*\ x_2(0) = 0$

$$f_2(1) = \min \begin{cases} .75f_3(1) = .525^* \text{ (0 hrs. on Eng.) } x_2(1) = 0 \\ .70f_3(0) = .630 \text{ (1 hr. on English)} \end{cases}$$

$$f_2(2) = \min \begin{cases} .75f_3(2) = {}^*.45 \text{ (0 hrs. on Eng.) } x_2(2) = 0 \\ .70f_3(1) = .49 \text{ (1 hr. on Eng.)} \\ .67f_3(0) = .603 \text{ (2 hrs. on Eng.)} \end{cases}$$

$$f_2(3) = \min \begin{cases} .75f_3(3) = .413^* \text{ (0 hrs. on Eng.) } x_2(3) = 0 \\ .70f_3(2) = .420 \text{ (1 hr. on Eng.)} \\ .67f_3(1) = .469 \text{ (2 hrs. on Eng.)} \\ .65f_3(0) = .585 \text{ (3 hrs. on Eng.)} \end{cases}$$

$$f_2(4) = \min \begin{cases} .75f_3(4) = .375^* \text{ (0 hrs. on Eng.) } x_2(4) = 0 \\ .70f_3(3) = .385 \text{ (1 hr. on Eng.)} \\ .67f_3(2) = .402 \text{ (2 hrs. on Eng.)} \\ .65f_3(1) = .455 \text{ (3 hrs. on Eng.)} \\ .62f_3(0) = .558 \text{ (4 hrs. on Eng.)} \end{cases}$$

$$f_1(4) = \min \begin{cases} .80f_2(4) = .30 \text{ (0 hrs. on French)} \\ .70f_2(3) = .289^* \text{(1 hr. on French) } x_1(4) = 1 \\ .65f_2(2) = .293 \text{ (2 hrs. on French)} \\ .62f_2(1) = .326 \text{ (3 hrs. on French)} \\ .60f_2(0) = .405 \text{ (4 hrs. on French)} \end{cases}$$

We spend $x_1(4) = 1$ hour studying French .$x_2(4\text{-}1) = 0$ hours studying English, and (of course!) 3 hours studying statistics. Her chance of passing at least one course is 1 - .289 = .711.

7. Define $f_t(w)$ to be the maximum net profit(revenues less costs) obtained from the steer during weeks t, t + 1,...10 given that the steer weighs w pounds at the beginning of week t. The key to the problem is to remember that revenues are only earned during week 10(because the steer is only sold once!) Then

$$f_{10}(w) = \max_p \{10g(w,p) - c(p)\}$$

where $0 \le p$.

Then for $t \le 9$

$$f_t(w) = \max_p \{-c(p) + f_{t+1}[g(w,p)]\}$$

Farmer Jones should work backwards until $f_1(w_0)$ has been computed.

Section 13.7

1. From EXCEL (file S13_7_1.xls) we find $f_1(0) = 8950$. From cell E17 we should produce 200 radios in month 1. Then we seek $f_2(0+200-200) = 6700$. From cell I16 we find that 600 radios should be produced during month 2. Then we seek $f_3(0+600-300) = 0$ and 0 radios should be produced during month 3.

Chapter 14 Solutions

1. Shortest path algorithm has two main steps. A labeling correcting step and a determining the minimum of the corrected labels.

Label Correcting Step: If (Label(I)+Distance(I,J)<=Label(J)) then Label(J) = label(I)+Distance(I,J) for all arcs out of node I.

Find Minimum of Corrected Labels:
Min = Large Number
For I = 1 to Number of Nodes
If (Label(I) is Not Permanent and Label(I) < Min)
Min = Label(I)
End For

Both the steps will be carried for n times for a problem with n nodes. The first step will be conducted for all the arcs in the graph. The total number of possible arcs if all the n nodes are connected to each other. That is number of arcs are (n-1)^2. Thus the comparison will be done at most (n-1)^2. Hence the complexity is O(n2).
The finding of the minimum valued node will be carried out at most n times, in which case the sink will be permanently labeled last. Each for execution will perform n operations. Thus this step has also has a complexity of O(n2). Since both the steps are of order (n2) the algorithm is of complexity O(n2).

classified as NP-complete.

5. The single machine scheduling problem with minimization tardiness with sequence dependent sequence time is same as TSP, with a different objective. Set initial temperature T to 2000; reduce temperature by 100 for the cooling schedule. Set number of iterations based on rule of thumb to 3 times number of jobs.

Step 1: Arrive at an initial sequence, preferably schedule in sync with order of due dates. This is known as early due date (EDD) schedule. Once the schedule is known, try starting the job first in the sequence to be scheduled immediately. Add change over time and processing time to determine the time job is available. If it is bigger than the du date, add the difference to the objective, otherwise the job is on or before the due date. After performing this we will arrive at the sum of tardiness for the schedule.

Step 2: For count = 1 to Number of iterations

Swap any two jobs. Calculate tardiness. Accept move if it is lesser objective, else use random number comparison to e(best objective–current objective/T) to decide accept or reject move.

End for

Step 3: Reduce T by 100, If T is not less than or equal to zero, repeat step 2 else report the best solution so far.

6. For minimizing early penalty and tardy penalty case, a simulated annealing, tabu search, or genetic search can be developed. In the case of simulated annealing and tabu search we will use exchange of jobs in the sequence as neighborhood moves or candidate list. In the genetic search, the initial population sequences will be randomly developed. The crossover and recombination function requires certain modifications to arrive at a feasible off spring. For example in 9 job problem, two strings chosen in selection step are:

```
1   3   4   7 | 9   8   2   6   5
5   4   7   9 | 2   8   6   3   1
```

Let the crossover position be 3.
The resulting strings after crossovers are:

```
1   3   4   9   2   8   6   3   1
5   4   7   7   9   8   2   6   5
```

These strings are infeasible, as the top one has scheduled job 1 and 3 twice without scheduling jobs 5 and 7. For the bottom string the situation is reversed, that is jobs 5 and 7 are scheduled twice, with jobs 1 and 3 not in the schedule. By swapping across the strings nearest twice-appearing jobs, we can arrive at a feasible schedule. In this case we will swap 1 and 5 in the first position, and 3 and 7 from position 2 in string 1 and position 3 in string 2. The result is:

```
5   7   4   9   2   8   6   3   1
1   4   3   7   9   8   2   6   5
```

The infeasible solution will have same number of mismatches between two siblings created by the same set of parents. Note that there are several possible schemes for doing such a trade.

Once a feasible schedule is developed by the procedures, evaluating the solution requires the following procedure:

The optimal idle time insertion technique of Garey, Tarjan, and Wilfong (1988) can be used with this heuristic once all jobs were placed and reordered. The basic idea behind this algorithm is jobs on a machine are grouped together (one or more groups), and each group may be moved earlier until either it can't be moved further or the number of tardy jobs is less than the number of non-tardy jobs in that block. The procedure is efficient, and it is reasonable easy to understand and implement. Another option is to solve a simple network flow problem for each string.

M. R. Garey, et al. One-processor scheduling with symmetric earliness and tardiness penalties. Mathematics of Operations Research 13 (2), (1988).

Chapter 15 Solutions

Section 15.2

2. See file S15_2_2.xls.

Section 15.3

2. See file S15_3_2.xls.

Section 15.6

1. See file S15_6_1.xls.

Section 15.8

2. See file S_15_8_1.xls.

Chapter 16 Solutions

1. The file S_16_3 contains the neural network predictions. The RSQ between the neural network predictions and actual cumulative area is .9999 and the average absolute error is .003.

	C	D	E	F	G	H
1	z	cum	prediction	abs dev		
2	-3.8	0.0001	0.0001167	1.666E-05		
3	-3.6	0.0002	0.0002715	7.151E-05	RSq	MAD
4	-3.4	0.0003	0.0005266	0.0002266	0.9998934	0.0027043
5	-3.2	0.0007	0.0009533	0.0002533		
6	-3	0.00014	0.0016704	0.0015304		
7	-2.8	0.0026	0.0028691	0.0002691		
8	-2.6	0.0047	0.0048496	0.0001496		
9	-2.4	0.0082	0.008076	0.000124		
10	-2.2	0.0139	0.0132584	0.0006416		
11	-2	0.0228	0.0214532	0.0013468		
12	-1.8	0.0359	0.0341381	0.0017619		
13	-1.6	0.0548	0.0531672	0.0016328		
14	-1.4	0.0808	0.080489	0.000311		
15	-1.2	0.1151	0.1175945	0.0024945		
16	-1	0.1587	0.1648943	0.0061943		
17	-0.8	0.2119	0.2214334	0.0095334		
18	-0.6	0.2743	0.2852549	0.0109549		
19	-0.4	0.3446	0.3542932	0.0096932		
20	-0.2	0.4207	0.4275225	0.0068225		
21	2.776E-16	0.5	0.5074494	0.0074494		
22	0.2	0.5793	0.5948592	0.0155592		
23	0.4	0.6554	0.6671261	0.0117261		
24	0.6	0.7257	0.7289626	0.0032626		
25	0.8	0.7881	0.7869286	0.0011714		
26	1	0.8413	0.8385728	0.0027272		
27	1.2	0.8849	0.8821195	0.0027805		
28	1.4	0.9192	0.9170489	0.0021511		
29	1.6	0.9452	0.9437638	0.0014362		
30	1.8	0.9641	0.9632987	0.0008013		
31	2	0.9772	0.9769974	0.0002026		
32	2.2	0.9861	0.9862264	0.0001264		
33	2.4	0.9918	0.9921857	0.0003857		
34	2.6	0.9953	0.9958306	0.0005306		
35	2.8	0.9974	0.9978956	0.0004956		
36	3	0.9986	0.998962	0.000362		
37	3.2	0.9993	0.9994758	0.0001758		
38	3.4	0.9997	0.999723	2.302E-05		
39	3.6	0.9998	0.9998499	4.986E-05		
40	3.8	0.9999	0.9999213	2.126E-05		

6. From file S16_6.xls you can see we found an RSQ of .9999 and MAD = .0029. This is virtually identical to what PREDICT found in Problem 3.